AF449241

THE SYNTAX AND SEMANTICS OF COMPLEX NOMINALS

THE SYNTAX AND SEMANTICS OF COMPLEX NOMINALS

Judith N. Levi

Department of Linguistics
Northwestern University
Evanston, Illinois

ACADEMIC PRESS

New York   San Francisco   London

A Subsidiary of Harcourt Brace Jovanovich, Publishers

ACADEMIC PRESS, INC.
111 Fifth Avenue, New York, New York 10003

United Kingdom Edition published by
ACADEMIC PRESS, INC. (LONDON) LTD.
24/28 Oval Road, London NW1   7DX

Library of Congress Cataloging in Publication Data

Levi, Judith N
    The syntax and semantics of complex nominals.

    Bibliography: p.
    Includes index.
    1.   English language——Noun phrase.   2.   English
language——Grammar, Generative.   I.   Title.
PE1201.L4   425   78–51236
ISBN 0–12–445150–0

To the memory of my mother
Anna Schapiro Levi

CONTENTS

1
INTRODUCTION

2
NOMINAL ORIGINS OF NONPREDICATING ADJECTIVES

8
SUMMARY AND CONCLUSIONS

FOREWORD

Professional linguists both do linguistics and talk about linguistics. The regrettable fact that recent linguistic publications reflect the latter activity so much more than the former makes Judy Levi's book on complex nominals an especially welcome addition to the literature. Existing textbooks in transformational grammar provide students with extensive training in the grinding of axes but little training in the use of already ground axes in the felling of trees. It is important that books such as Levi's be widely read, since only through seeing broad factual areas subjected to detailed analysis that pays attention to but does not wallow in matters of linguistic theory can one gain the perspective needed to judge the relative importance of different problems that arise in linguistics, the correctness of purported facts and generalizations that have figured in theoretical arguments, and the wisdom or foolishness of linguists' evaluations as to what approaches and considerations deserve to be taken seriously and which ones can be dismissed out of hand. If reading a typical textbook of transformational grammar can be compared to watching an automobile commercial, reading Levi's book can be compared to test-driving a car on rugged terrain.

That terrain is not only rugged but also rather lonely, once one gets away from the tourist center that has developed around the site of a much discussed and frequently reenacted battle. Levi's book should have the salutary effect of reducing the disparity between the attention that has been lavished on nominalizations (the site of the celebrated battle) and the comparatively little attention that has been given to the vast territory of complex nominals, of which nominalizations form only a corner. Levi has shown complex nominals (including nominalizations) to be largely regular in their syntax and semantics, with their irregularity mostly in the realm of morphology (e.g., whether something with the meaning of a compound noun can be expressed in Adjective + Noun form depends on whether the first member of the compound has an adjectival form, which is largely unpredictable) and much of their alleged irregularity being only apparent (e.g., complex nominals are generally ambiguous as to which of a limited set of predicates has been deleted, but real world knowledge and factors of cooperativity between speaker and addressee cause the ambiguity to be resolved in different ways in different cases). Levi's analysis is done within a fairly specific theoretical framework (generative semantics with global rules and surface target structures) but provides an identification of what is regular and what is irregular in different kinds of complex nominals that will be invaluable to anyone concerned with the analysis of these constructions or with the theoretical issues that they raise.

Two virtues of this book, over and above the importance of its content and the clarity of its exposition, lead me to urge that it be read not only by professional linguists interested in the many important points of the structure of English and of linguistic theory that it deals with but also by neophyte linguists, say, students in the second semester of a one-year sequence on transformational syntax. First, there is the gratifying degree of detail that Levi goes into in stating the facts, in presenting derivations of examples, and in describing the rules that are involved in those derivations. This is a welcome departure from the tradition of fragmentary derivations and grossly underspecified rules that has been typical of the generative semantic literature. Second, there is Levi's policy of presenting her rules not in terms of the widely accepted notations that have been developed for formulating transformations but by exhibiting the configuration of items that the transformation affects and its effect on that structure, together with an informal statement of the restrictions to which the application of the rule is subject. Levi's presentation is more explicit than one in terms of the "standard" notation would be (since the "standard" notation often leaves unclear the constituent structure of the output) and frees her and her readers from any need to digress into questions that the choice of a notation is often alleged to provide an answer to, for example, the question of what is a possible grammar of a natural language. Those are important questions, but one is not obliged to answer them before being permitted to do any linguistics; indeed, one is in no position to answer them until one has done quite a lot of linguistics.[1] Levi's book thus provides welcome relief from the common tendency among linguists to identify precision with formalism and to describe the metalanguage in greater detail than the language.

I would like to close with an expression of personal pleasure and satisfaction at the publication of this book. Much of the material contained here originated in a dissertation done under my supervision that had the rare distinction of starting out as a rather ambitious project and ending up by delivering even more than the author had promised in her rash thesis proposal. I am grateful to Judy Levi for demonstrating that, at least once in a while, audacious claims about language can survive a detailed and critical examination of the facts, and for taking some of my ideas about language more seriously than I myself did.

James D. McCawley
University of Chicago

---

[1] See my papers "Some ideas not to live by" (*Die neueren Sprachen* 75, 1976, pp. 151–165) and "Acquisition models as models of acquisition" (R. Fasold and R. Shuy, eds., *Studies in Language Variation*, Washington: Georgetown University Press, 1977, pp. 51–64) for arguments against the conception of language acquisition that provides much of the basis for the prevalent attitude of transformational grammarians toward notation, and "The role of notation in generative phonology" (M. Gross, M. Halle, and M.-P. Schutzenberger, eds., *The formal analysis of natural language*, The Hague: Mouton, 1973, pp. 51–62) for critical discussion of arguments that transformational grammarians have offered for choices among notational systems.

PREFACE

Oddly enough, I did not set out to study complex nominals. My research began instead with an intriguing set of "attributive-only" or "nonpredicating" adjectives (as in *electrical engineer* and *lunar exploration*), whose somewhat perplexing properties had come to my attention through Bolinger's insightful presentation in a 1967 article in *Lingua*. Once having recognized, however, that these adjectives function primarily as noun substitutes within complex nominals, I was led ineluctably by my own hypothesis to enter the broader and ultimately more significant area of grammar which comprises both "nominal compounds" and nominalization formations of various types. These constructions, which I have grouped together under the heading of "complex nominals," formed the major subject of my dissertation research, out of which the present study has developed.

My major purpose in writing this book has been to explore in depth the syntactic and semantic regularities I believe to be characteristic of nonlexicalized complex nominals in English and, quite possibly, in all natural languages. To the certain relief of some and the probable chagrin of others, my perspective throughout has been much more data-oriented than metatheoretical in nature. As a result, although my analysis is presented within a generative semantic framework, it has been my hope and intention that my work will be useful to theoretical linguists across the whole spectrum of contemporary linguistic thought, as well as to psycholinguists, applied linguists, and, indeed, many nonlinguists whose work requires an understanding of the nature of complex nominals. In this latter group I would include developmental psycholinguists and cognitive scientists, specialists in artificial intelligence and natural language processing, lexicographers and philosophers of language, and teachers of English as a second language and foreign languages in general.

While this book makes available to a larger audience the most important portions of my earlier work, those already familiar with my dissertation research will find much here that is new, as well as significant revisions in the theoretical core common to both. For example, the scope of the theory has been carefully redefined to make a narrower but more accurate range of predictions, the set of Recoverably Deletable Predicates has been expanded from seven to nine members, and a major new analysis of nominalization types in English has been incorporated into the overall theory. In addition, I have attempted to suggest some of the ways in which my own work may relate to a number of new issues of theoretical concern, including the function of complex nominals in discourse, the interaction of pragmatic principles with syntactic and semantic regularities, and the relationship of productive word formation processes to our theories of syntax and morphology.

There are, on the other hand, some tasks and topics I have chosen not to include in the book. For example, I have not attempted any substantive metatheoretical comparisons in these pages, not only because such discussion would exceed the space limitations on this volume but also because the absence of any competing analyses of comparable scope and focus precludes any such attempt at the present. I have also avoided the thankless task of trying to present my analysis in such a way as to please linguists of **all** theoretical persuasions, although it remains possible that my innovations may please none. In particular, I have not formulated my rules in the classical mode of Structural Description and Structural Change because I have often found such linear formulations to be less than fully informative, if not potentially misleading; instead, I have presented each of the crucial rules of Chapters 4 and 5 in the form of detailed tree structures accompanied by specific conditions and explicit commentary, in the belief that this format will convey more clearly and completely the actual content of my proposals.

In summary, my primary intention has been to construct as coherent and comprehensive an analysis of the syntax and semantics of complex nominals as I could, within the theory and constraints of generative semantics as I have understood it. In those cases where I was obliged to "patch up" generative semantic theory in somewhat unorthodox ways, I have tried to present an honest account of just what the empirical and theoretical problems were, regardless of whether I myself was satisfied with my own solution. Although the raggedness of these patches may be due in some cases to my own ineptitude, in others I have serious doubts as to whether any of our present theories could weave **all** of the relevant data into a smoother or more regular fabric.

When I began my research some five years ago, I could not have anticipated just how rich and rewarding this area of the grammar would prove to be for linguistic exploration and discovery. Not only has it furnished an unending progression of puzzles and problems to wrestle with and worry over, but it has suggested a splendid array of intellectual challenges for future research in a variety of disciplines. Among these must be counted such important issues as the "location" of word formation processes in regard to the alleged boundary between syntax and morphology, the manner in which children learn to master and manipulate complex nominals, the degree to which the principles governing complex nominal formation in English are reflections of universal constraints on such processes, and the ways in which syntactic and semantic regularities of complex nominals may be exploited in the many applications of artificial intelligence and natural language processing. It is my hope that what I have done in this book, as well as what I have failed to do, will stimulate further research in all these areas so that we may continue to increase our understanding of complex nominals and their significance in natural language.

ACKNOWLEDGMENTS

It is with pleasure that I address myself to thanking those who have contributed in important if diverse ways to the writing of this book. As I look back on the years of study and research that have culminated in the publication of this volume, I recognize that I have been extremely fortunate in being part of a scholarly community that has provided me with the excitement of intellectual curiosity and acuity, an atmosphere of mutual respect and encouragement on both personal and professional planes, and a network of warm and supportive friendships persisting despite separations in time and space. The very many individuals who have contributed to creating this community and to making me feel a part of it are indeed too numerous to name here. Within this community, however, are a number of linguists and friends who unquestionably deserve individual mention.

I shall begin where the research for this book began: with my dissertation and my dissertation committee at the University of Chicago. Gene Gragg, James D. McCawley, and Jerrold M. Sadock not only contributed an appalling number of hours to reading and commenting on two 400-page drafts, but—far more significantly—provided me with superb examples of what it means to combine intellectual excellence, scholarly cooperation, patient teaching, and good-humored collegiality. Their skillful guidance and rigorous instruction gave me the firm foundation in theoretical linguistics without which this book could not have been written.

In earlier stages of my research, I profited from correspondence with Dwight Bolinger, Robert B. Lees, and Paul M. Postal, all of whom I wish to thank for their willingness to help and encourage a novice in the field. I also learned a great deal through correspondence with Pamela Ann Downing and Frederick J. Newmeyer, for our theoretical disagreements obliged me to sharpen my thinking and reexamine my original claims; I am thus grateful to both of them for the intellectual stimulation and constructive dialogue which contributed in important ways to the development of this book.

In the area of crosslinguistic work, I would like to thank Inge Bartning, Joseph C. Chryst, Müşerref Dede, and Yael Ziv for their interest in discussing with me the application of my theory to French, Japanese, Turkish, and Modern Hebrew, respectively; needless to say, they are absolved of any responsibility for whatever brash conclusions I may have drawn subsequently concerning the place of complex nominals in universal grammar.

In the later stages of my writing, Eric P. Hamp, Janet Cohen Jenkins, James D. McCawley, Jerrold M. Sadock, and Ellen Schauber were always gracious to me when I would corner them with a theoretical puzzle to solve or a set of data to worry about; I appreciate their tolerance of my writer's symptoms of anxiety as much as the real analytic talent which they brought to bear on the problems at hand.

There is no question in my mind, however, that the person to whom I owe the largest debt of intellectual and personal gratitude for her help on this book is Georgia M. Green, who read and commented at length on an earlier version of the manuscript. A superb critic, Georgia has a kind of x-ray vision which seems to penetrate instantly to whatever unexamined assumptions, internal contradictions, and imprecise thinking may be lurking behind a certain word or phrase. Her queries and comments on matters both stylistic and substantive were invaluable, not least because they consistently reflected the tacit assumption that only the highest standards of accuracy, clarity, and explicitness would suffice. As a result, Georgia taught me much and kept me honest. For all that she has given me—in time, insight, inspiration, and intellectual challenge—I am deeply grateful.

I would also like to thank Joan Chelnick and Janet Cohen Jenkins, who made my life easier by their patient typing of draft after draft, and Mary S. Blaisdell who assisted in numerous ways with the physical preparation of the manuscript. In addition, my editors at Academic Press all helped to make our collaboration a truly congenial and productive enterprise; I thank them most sincerely.

　　　　　　　　　　　　　　　　　　　　　

Finally, I am grateful to the Chicago Linguistic Society and to North-Holland Publishing Company for permission to include in this book material based on articles of mine which they had previously published.

LIST OF ABBREVIATIONS

adj, Adj	adjective
adv, Adv	adverb
ARG	argument (in Logical Structure)
CN	complex nominal
Comp Adj Formation	Compound Adjective Formation
CSC	Coordinate Structure Constraint (see Ross 1967)
DET	determiner
GEN	genitive
IS	Intermediate Structure
LS	Logical Structure
Morph Adj	Morphological Adjectivalization
N	noun
N_1	prenominal modifier (nom adj or N) in a CN
N_2	head noun in a CN
NC	nominal compound (= compound noun)
NEG	negative
NOM	nominalization
nom adj	nominal (nonpredicating) adjective
NOM CN	CN derived by predicate nominalization
nom nonpred adj	nominal nonpredicating adjective
nonpred adj	nonpredicating adjective

NP	noun phrase
PA	pseudo-adjective (see Postal 1969, 1972)
POSS	possessive
PP	prepositional phrase
PPA	proper pseudo-adjective (see Postal 1969, 1972)
pred, Pred	predicate
PRED	predicate (in Logical Structure)
prep, Prep	preposition
Prod	Product (Nominalization)
RDP	Recoverably Delectable Predicate
RDP Deletion	Recoverable Predicate Deletion
Rel Cl	relative clause
Rel Pro	relative pronoun
S	sentence
SS	Surface Structure
Subj Gen	Subjective Genitive (Formation)
SUH	Strong Universals Hypothesis
V	verb
WH–*be* Deletion	Relative Clause Reduction

1

INTRODUCTION

1.1 The Data

This book is devoted to the study of the syntactic and semantic properties of "complex nominals" in English. This term is used to encompass the three partially overlapping sets of expressions which are illustrated in (1.1)–(1.3); these have generally been called "nominal compounds," "nominalizations," and "noun phrases with nonpredicating adjectives," respectively.

(1.1)   *apple cake*   *deficiency disease*
        time bomb   *autumn rains*
        doghouse   *nicotine fit*
        windmill   *color television*
        daisy chain   *surface tension*

(1.2)   *Markovian solution*   *film producer*
        American attack   *city planner*
        presidential refusal   *dream analysis*
        musical criticism   *quantifier lowering*
        constitutional amendment   *metal detection*

(1.3)   *electric clock*   *musical clock*
          electric shock   *musical criticism*
          electrical engineering   *musical interlude*
          electrical conductor   *musical comedy*
          electrical outlet   *musical talent*

The most superficial examination of these data reveals that what these expressions have in common is a head noun preceded by a modifying element which in some cases is a noun, in others what appears to be an adjective. It is the purpose of this book to explain the many other features that these expressions share which are not so apparent at first glance.

The three subsets of "complex nominals" shown in (1.1)–(1.3) have received varying degrees of attention from linguists in the past. At one end of the scale of popularity are the **nominal compounds** shown in (1.1), which have been intensively scrutinized over a period of centuries by linguists studying such diverse languages as English, Turkish, Hebrew, Chinese, Sanskrit, and German. Among the most notable studies of English compounds are those found in Jespersen 1942, Koziol 1937, Lees 1960, and Marchand 1969.

Nominalizations like those in (1.2) have also been frequently analyzed, sometimes in their own right and sometimes as a subgroup within a compound typology. In recent American linguistics, the analysis of nominalizations has played a prominent role in the theoretical controversy between the schools of "lexicalism" and "generative semantics," as may be seen by examining the crucial analysis of Chomsky 1970 on the one hand, and critical responses such as McCawley 1975, Newmeyer 1971, 1975, 1976, and Ross 1974a on the other. Even these studies, however, have provided only preliminary insights into the nature of the nominalization process, so that our understanding of expressions like those in (1.2) remains far from complete.

Probably the least explored subset of complex nominals is that represented by the forms in (1.3). Indeed, it was not until 1974 that these forms received a name of their own (**nonpredicate NPs**, in Levi 1974:403), although the adjectives alone had been identified as a distinct set in earlier studies, where they were variously called "pseudoadjectives," "attributive-only adjectives," "denominal adjectives," "transposed adjectives," and "denominal nonpredicate adjectives."[1] It was, in fact, this author's investigation of the intriguing properties of this group of adjectives which provided the foundation for the present study of complex

---

[1] These terms are found in Postal 1969, 1972, Bolinger 1967, Ljung 1970, Marchand 1969, and Levi 1973, respectively.

nominals. Let us consider for a moment just why these adjectives merit special attention.

One of the most striking features of these adjectives is that they may not be used in the predicate (post-copula) position in which most other adjectives regularly appear; this may be seen by contrasting the grammaticality and semantic equivalence of the expressions in (1.4) with those in (1.5).

(1.4)   a. *a beautiful princess*   b. *a princess who is beautiful*
           a logical conclusion   *a conclusion which is logical*
           a clever engineer   *an engineer who is clever*
           an efficient generator   *a generator which is efficient*

(1.5)   a. *a rural policeman*   b. **a policeman who is rural*
           a logical fallacy   **?a fallacy which is logical*
           an electrical engineer   **an engineer who is electrical*
           a solar generator   **a generator which is solar*

Bolinger (1967) seems to have been the first to realize that this distinction between the "normal" adjectives in (1.4) and the apparently "abnormal" adjectives in (1.5) constituted counterevidence to the oft-repeated claim of early transformational work that **all** prenominal adjectives in English are derived from predicate position in relative clauses.[2] In his 1967 article, Bolinger demonstrates that this claim makes the incorrect prediction that the expressions in (1.5a) must be derived from intermediate stages corresponding to the expressions shown in (1.5b). Such a derivation cannot be correct, however, since forms like those in (1.5b) either are semantically ill-formed (as is true, for example, of **a policeman who is rural*) or have a well-formed reading which is nonsynonymous with the counterpart form of (1.5a). The latter case is illustrated by these two expressions from (1.5): *a logical fallacy* (in the sense of 'a fallacy in logic or logical reasoning') and *a fallacy which is logical* (in the sense of 'a fallacy which conforms to the rules of logic'). Under the assumptions of generative semantic theory, we must rule out deriving the former from the latter (or, more precisely, deriving both from an identical source) on the grounds that nonsynonymous expressions may not be derived from the same underlying structure.

Another puzzling feature of the adjectives in (1.3) (to be called hereafter **nonpredicating adjectives,** or **nonpred adjs**) is the fact that their meaning appears to change, depending on the head noun that they modify. In (1.3), for example, we understand *musical clock* to mean 'a

[2] See, for example, Burt 1971:81ff. or Jacobs and Rosenbaum 1968:204–205.

clock that produces music'; yet we also know that *musical criticism* could hardly be construed as 'criticism that produces music', but must mean instead 'criticism **of** music', where *music* functions as the object of a nominalized verb. However, this latter analysis fails to apply to any of the remaining examples with *musical* (i.e., *interlude, comedy,* and *talent*), each of which has its own appropriate—and distinct—paraphrase. Finally, it is significant that none of these expressions uses *musical* in the sense of the normal predicating adjective shown in (1.6), namely, with a meaning of 'like music, melodious, euphonious':

(1.6)   a. *She has a musical voice.*
         b. *Her voice is very musical.*

These data force us to ask just how many adjectives English can have with the homophonous form of *musical*, and, as a corollary question, whether there are any limits to the apparent polysemy of nonpredicating adjectives in general.

Still another distinctive characteristic of these adjectives is that they regularly appear in positions where we might otherwise expect nouns; that is, they form complex nominals that are strikingly parallel in semantics and syntax to the noun–noun collocations that have long been identified as **compounds.** In (1.7) we see examples of semantically parallel expressions, while (1.8) shows examples of fully synonymous pairs.

(1.7)   a. *corporate lawyer*   b. *tax lawyer*
            sanitary engineer   *mining engineer*
            electrical shock   *future shock*
            molecular chains   *daisy chains*
            parental prerogative   *student prerogative*

(1.8)   a. *linguistic difficulties* b. *language difficulties*
            industrial output   *industry output*
            dramatic criticism   *drama criticism*
            oceanic winds   *ocean winds*
            atomic bomb   *atom bomb*

The fact that nonpredicating adjectives appear to function just like nouns in expressions such as those in (1.7) and (1.8) strongly suggests the hypothesis that such adjectives are derived from underlying nouns. Support for this hypothesis will be provided in detail in Chapter 2, where it will be shown that such a theory of nominal origins can explain the otherwise anomalous syntactic and semantic behavior of this group of adjectives. Since, however, this theory makes the claim that the data on nonpredicating adjectives are only superficially distinct from the data on both compounds and nominalizations, we are led inexorably to the conclusion that an analysis of any one of these sets of data can only be made as part of a comprehensive analysis of all three. The present work is, then, an attempt to provide just such an analysis.

1.2 General Purposes and Hypotheses

The principal purpose of this book is the exploration of the syntactic and semantic properties of **complex nominals** (**CN**s) in English, and the elaboration of detailed derivations for these forms within a theory of generative semantics. The book also represents an attempt to incorporate into a grammar of English a model of the **productive** aspects of complex nominal formation, and to analyze in what ways these must be distinguished from more idiosyncratic aspects of the same forms. The conclusions of this study of English will then be placed in the broader context of universal grammar, in order to understand what elements of the theory are likely to reflect universal principles of CN formation. A number of other ramifications of the theory will also be discussed, including its relevance to lexicographic treatments of CNs and the interplay of syntactic, semantic, and pragmatic factors in our usage of CNs.

As indicated earlier, the first hypothesis to be taken up will be that which claims that nonpredicating adjectives like those appearing in (1.7) and (1.8) are derived from underlying nouns; this hypothesis correctly predicts that these adjectives will manifest semantic and syntactic characteristics of nouns, despite their surface morphological difference. Arguments in support of this hypothesis will then lead to the claim that there exists a set of structures, termed **complex nominals,** which includes the three data sets shown in (1.1)–(1.3).

Evidence will be presented to show that CNs are dominated by a node label of N on the surface, while their more remote source structures are dominated by an NP node, whose daughter nodes consist of a head NP and an S node, in either a relative clause or NP complement construction. Because of the recursive nature of CN formation, a CN may be as short as the standard N–N compound (e.g., *apple cake*) or as long as such expressions as *horseback riding school cafeteria breakfast menu substitution list*. Modifying nouns in certain structurally defined positions (to be specified later) will be subject to an optional rule of adjectivalization, so that the number of nonpredicating adjectives in CNs will vary from zero (as in the two previous examples) on up (cf. *polar bear, tropical tidal wave,* or *chemical engineering instructional manual editorial staff*).

One of the most important claims of the present work is that all CNs must be derived by just two syntactic processes: predicate nominalization and predicate deletion. The analysis of CNs derived by the latter process will entail the proposal of a set of Recoverably Deletable Predicates (RDPs) representing the only semantic relations which can underlie CNs. This set, whose members are small in number, specifiable, and probably universal, consists of these nine predicates: CAUSE, HAVE, MAKE, USE, BE, IN, FOR, FROM, and ABOUT. The fact that a given CN may be derived by the deletion of any one of these predicates, or by the process of nominalization, means that CNs typically exhibit multiple ambiguity. This ambiguity is, however, reduced to manageable proportions in actual discourse by semantic, lexical, and pragmatic clues.

1.3 Scope Restrictions

It is important to specify at the outset that the scope of this study is restricted to **endocentric** complex nominals, that is, those CNs whose referents constitute a subset of the set of objects denoted by the head noun. This will eliminate the three groups of exocentric CNs of which I am aware:

1. Those which seem to be metaphorical names, such as *ladyfinger* (for a type of pastry), *tobaccobox* (for a sunfish), *silverfish* (for an insect), or *foxglove* (for a flower);
2. Those which describe their referent synecdochically, by using a part to represent the whole, as in *peg leg, blockhead, birdbrain,* or *eagle-eyes* to describe people; or *razorback, glasseye, hammerhead,* and *cottontail* to describe animals;
3. Those which constitute coordinate structures such that neither noun may be taken as head, as in *speaker–listener, participant–observer, player–coach, secretary–treasurer, screwdriver–hammer, sofa–bed,* or *library–guestroom*.[3]

---

[3] It is interesting to observe that the semantic relation between the surface form and the unexpressed head noun in these exocentric examples is BE in the case of both the metaphorical names and the coordinate CNs, and HAVE in the case of the synecdochical names; thus, *foxglove* means 'flower which IS [metaphorically] a fox's glove', *speaker–listener* means 'person who IS (both) a speaker and a listener', and *birdbrain* means 'person who HAS the brain of a bird'. This fact is significant in that both BE and HAVE belong to the set of predicates that I claim have a special status in CN formation as well as in other areas of the grammar. It is therefore quite likely that exocentric CNs represent a special case of my theory, involving the formation of a CN by regular rules followed by the loss of the head noun by "beheading" (a notoriously idiosyncratic process discussed in Borkin 1972 and Jenkins 1976). I shall not, however, pursue this notion further in this work.

(Note, however, that the **literal** readings of the metaphorical and synecdochical types do fall within the scope of my theory; most often, these readings are based on either BE or HAVE Deletion, as in *hammerhead* and *ladyfinger*, respectively.)

Also to be excluded from consideration are those proper nouns that resemble CNs in form but whose first element is used primarily to **name** a single, definite referent; examples of this type, which usually denote specific places or businesses, include *Istanbul Hotel, Parisian Café, Kennedy Library, Rockefeller University, Atlas Enterprises,* and *Sheridan Road*. These are excluded on the grounds that the principles according to which they are formed are distinct from those of normal CNs; in particular, the initial element could be replaced by any number of lexical items (or, for that matter, by nonlexemes or even plain numbers) and still denote the same object, a condition which is certainly not true of complex nominals in general.[4] Since these expressions are thus not **derived** by systematic grammatical processes, but are instead **chosen** or **invented** for reasons which are generally extralinguistic (e.g., cultural connotations, esthetic principles, financial obligations), they do not belong to the present study.

A third set of data that is outside the scope of this work consists of nominal expressions whose prenominal modifiers may appropriately be called nonpredicating adjectives, but which must be derived from underlying adverbs rather than underlying nouns. Some examples from this group are shown in (1.9).

(1.9)

potential enemy	*occasional visitor*
former roommate	*alleged attacks*
puzzled frown	*putative solution*
disgruntled comment	*eventual compromise*
early riser	*individual decisions*
heavy smoker	*future dependents*

As the reader can easily verify, these expressions, like those shown earlier in (1.3), cannot have been derived from a relative clause construction in which the same adjective appears in predicate position. That these adjectives are derived from adverbs rather than from nouns, however, is suggested by paraphrase possibilities like those in (1.10).

---

[4] Thus, there is nothing systematic about the fact that the street on which my office is located is referred to by the expression *Sheridan Road*; it could just as well have been called *Istanbul Road, Atlas Road, Judy Levi Road, Q Road,* or *Road 379*. In contrast, the arbitrary substitution of other nouns for the first element in a normal CN would lead to solipsistic (and dysfunctional) nonsense. This distinction follows from the name-creating function served by CNs in natural language (a topic to be taken up in Section 3.4).

(1.10)   a. *They are all* $\begin{Bmatrix} potential\ enemies \\ potentially\ enemies \end{Bmatrix}$.

b. *She* $\begin{Bmatrix} is\ a\ former\ roommate \\ was\ formerly\ a\ roommate \end{Bmatrix}$ *of Sue's.*

c. *His parents* $\begin{Bmatrix} are\ both\ heavy\ smokers \\ both\ smoke\ heavily \end{Bmatrix}$.

d. *I question the adequacy of* $\begin{Bmatrix} the\ putative\ solution \\ what\ is\ putatively\ the\ solution \end{Bmatrix}$.

Since the adjectives in question are not derivable from nouns, they lack the syntactic and semantic parallels with "compounds" that are crucial here, and hence will be excluded from the present analysis.[5]

Finally, it must be emphasized that the theory to be elaborated in the pages to follow is a theory of the **productive** processes that generate complex nominals; this means that the data for which my theory can provide an explanation are limited to nonlexicalized, nonspecialized, nonidiomatic, and (for the most part) nonmetaphorical forms. There has been much confusion in the literature stemming from the failure of both authors and critics to distinguish among these different sets of data, yet the distinction is a critical one. The problem is well stated by Motsch (1970):

> It is necessary to distinguish between the process of forming compounds and formations already existing in the lexicon, that is: between the grammatical rules which create potential compounds, and lexicalized forms which have been taken into the lexicon of a language as expressions and which can be systematically analyzed only to a limited extent. Lexicalized forms may have a long history, in the course of which properties which must be characterized as idiosyncratic may be acquired. With this distinction presupposed, we will investigate only the grammatical properties of potential compounds [p. 210; translation mine—JNL].

Part of the confusion around this issue stems from the fact that although CNs are **potentially** ambiguous over a number of readings, most CNs which are in common use have what has been called an "accepted reading," an "institutionalized referent," or a "most probable reading," which virtually eliminates ambiguity in regular discourse. Thus, the CN *horse doctor* is normally used with the meaning shown in (1.11a), although it could have any of the other meanings in (1.11); similarly, the CN *steam iron* has an "institutionalized referent" described

---

[5] These forms are discussed in more detail in Levi 1975, Chapter 7 and in Berman 1974, Section 3.1. Relatively limited subsets of these adjectives are also examined in Bach 1968, Bolinger 1967, Dixon 1970, Givón 1970, Hall 1973, Ljung 1970, and Vendler 1968.

by (1.12a), although it too could be ambiguous over a number of readings, including all those shown in (1.12).

(1.11)	a. 'doctor for horses'	(cf. *tree doctor*)
	b. 'doctor who is a horse'	(cf. *woman doctor*)
	c. 'doctor that has horses'	(cf. *peg leg doctor*)
	d. 'doctor that uses horses'	(cf. *voodoo doctor*)

(1.12)	a. 'iron that uses steam [to function]'	(cf. *electric iron*)
	b. 'iron made up of steam'	(cf. *copper iron*)
	c. 'iron for (ironing) steam'	(cf. *trouser iron*)
	d. 'iron which is (located) in steam'	(cf. *attic iron*)

It will be the purpose of this study to specify **all** the possible meanings that a given CN can have, rather than just the "accepted" or "institutionalized" readings.[6]

Nonetheless, the multiple readings that my theory will associate with a given surface CN will **not** include metaphorical, lexicalized, or idiomatic meanings. **Metaphorical** usage of a CN may involve extending the sphere of reference of the CN as a whole, or creating a CN in which only one member is intended metaphorically. Examples of the former appear in (1.13), while (1.14) and (1.15) illustrate CNs in which just the first or just the second noun is used metaphorically.

(1.13)	a. *icebox* = 'place as cold as an icebox'
	b. *windbag* = 'person who talks too much'
	c. *bottleneck* = 'obstruction of some process or flow [as in traffic or legislative procedures]'
	d. *snake eyes* = 'throw of two single dots on two dice'
	e. *coffin nails* = 'cigarettes'
	f. *beehive* = 'a place as busy as a beehive'

| (1.14) | *hairpin turn, beehive hairdo, handlebar mustache, virgin forest, queen bee, parent organization, pet theory*[7] |

---

[6] This is true despite the fact that familiar CNs will usually be listed here (and in the Appendix) under their most common interpretation (e.g., *steam iron* under USE, *horse doctor* under FOR); by thus capitalizing on our knowledge of the "institutionalized referent" of each CN, we avoid the necessity of glossing each form. The justification for such a shortcut must, however, be understood as solely pragmatic rather than theoretically significant in any way.

[7] In my theory, the CNs in (1.14) are analyzed as having an equational predicate BE deleted; for example, *handlebar mustache* is derived from the same source as *mustache which is a handlebar*. Although other linguists have proposed analyses in which such CNs would come from an underlying source of the form *x which is* LIKE *y*, I argue in Section 4.1.4 that extended uses of this sort can be better explained by pragmatic principles of interpretation than by the deletion of a predicate equivalent to LIKE.

(1.15)   *cannon fodder, jailbird, road hog, sand dollar, sea horse, brainstorm, wallflower*

In all these cases, my theory will be able to predict the **literal** but not the metaphorical senses of these expressions.[8] This is, however, not due to any descriptive inadequacy of my theory but rather to the nature of metaphor itself, which derives its expressive force precisely from its creative, and hence idiosyncratic, associations and substitutions.

Lexicalized meanings of CNs will also not be predicted by my theory. "Lexicalized meaning" refers to that semantic information idiosyncratically accruing to a particular surface form which (*a*) does **not** represent regular grammatical processes; (*b*) must be learned on an item-by-item basis; and hence (*c*) must be included in a separate entry for that CN in the lexicon of the language in question. For example, my theory will predict a number of possible readings for each of the CNs in (1.16)– (1.19), including the "accepted" reading shown in each case as (a); it will not, however, predict the lexical specialization, indicated in (b), that has been added to the meaning of each of these CNs. I also claim that **no** grammatical theory could make such predictions.[9]

(1.16)   *ball park:*   a. 'park for ball'
                             b. 'park or stadium designed for people to play baseball in [rather than football, basketball, or handball]'

(1.17)   *shell shock:*   a. 'shock caused by a shell or shells'
                             b. 'psychological trauma suffered by soldiers due to battlefield experiences'

(1.18)   *eggplant:*   a. 'plant which is (like) an egg'
                             b. 'purple ovoid fruit of the plant *Solanum melongena*'

(1.19)   *manual labor:*   a. 'labor using the hands'
                             b. 'strenuous physical labor, generally not requiring verbal or abstract mental skills'

[8] There is in fact no conceivable way for a theory to predict the idiosyncratic associations of meaning and form shown in (1.13), unless by some very general principle that says, "Any term x may be used to denote any object y in such a way as to include the meaning 'y is like x in some [not overtly specified] way' in the lexical entry of x as a name for y." See also Reddy 1969 in this regard.

[9] A similar point is made by Motsch (1970:222) in his study of N–N compounds in English (q.v.). This does not mean, however, that it is an easy matter to distinguish between lexicalized and nonlexicalized CNs; for a discussion of some of the issues involved, see Sections 6.2 and 7.3.

An objection might be raised at this point that I am simply excluding a priori from the scope of my theory those CNs for which it fails to make correct predictions. The answer to this objection is that an insistence on maintaining the distinction between nonlexicalized and lexicalized CNs entails making certain empirical predictions, which can then be verified or disproved. These predictions include the following:

1. Both native speakers and foreigners learning English will have to learn the lexicalized senses (or referents) of CNs on an individual basis, and will be likely to make errors in matching term and referent until such learning has taken place.[10]

2. Lexicalized CNs will "lose something in translation"; that is, they either will have no word-for-word equivalent (e.g., *eggplant* becomes *aubergine* in French, not **plante* [*d'*]*oeuf*) or will define a different set of referents than the superficially corresponding foreign term. (For example, *manual labor* and *mother wit*, translated literally into Hebrew, give us *avoda yadanit* 'handwork' and *šninut imahit* 'maternal cleverness', respectively; the direct translation thus preserves the literal but not the specialized senses of the English forms.)

3. Somewhat similarly, the substitution of synonyms in English for each of the components of a lexicalized CN will not produce a second CN with the same set of referents as the first; thus, the lexicalized CNs *grammar school*, *venereal disease*, and *residential treatment* do not refer to the same entities as those likely to be named by *language-structure academy*, *Venus' illness*, and *home therapy*, respectively. In contrast, the nonlexicalized CNs *canine companion*, *marine life*, and *water bugs* may readily be replaced by such close synonyms as *dog escort*, *sea biology*, and *aquatic insects*, respectively.

4. Lexicalized CNs will be less suitable than nonlexicalized CNs (if not totally unsuitable) as models for additional examples, a difference that follows from the fact that the former are themselves not simply the output of a general rule (e.g., the 'strenuous, nonverbal' components of the lexicalized meaning of *manual labor* cannot be used to form an analogous CN *manual sports* in the sense of 'strenuous, nonverbal sports activities').

A distinction can also be made between "lexicalized" and "idiomatic" meanings of a CN, although this distinction is not crucial to the present

[10] For example, a foreigner asked to meet some American friends "at the ball park" could understandably end up—alone—at a football field or outdoor handball court, unless he had specifically learned that Americans use *ball park* to refer to playing areas for baseball only. Similarly, a child asked to pick up an eggplant at the store might just as easily (and reasonably) come home with an onion or a mango if she had not yet learned the actual referent of this term.

study since both fall outside its scope. I will use the term "lexicalized" to refer to elements of the meaning of a CN that appear to be added on to a more literal meaning which **is** predicted by my theory; thus, the information shown in the (b) entries of examples (1.16)–(1.19) includes the general senses given in (a), but represents a semantic specialization which narrows down the latter in reference. In contrast, what I will call "idiomatic" readings of CNs are readings for which the component nouns of the CN are more or less irrelevant. Totally idiomatic CNs (that is, those in which neither noun plays any role in the semantic interpretation of the CN) are shown in (1.20), along with their idiomatic readings:[11]

(1.20)　　a. *duck soup* = 'a snap, very easy'
　　　　　b. *honeymoon* = 'trip taken by newlyweds to celebrate their
　　　　　　　　　marriage'
　　　　　c. *fiddlesticks, horsefeathers, bullshit* = 'nonsense'
　　　　　d. *swan song* = 'last work of an artist, shortly before death'
　　　　　e. *soap opera* = 'radio or TV show depicting dramatic events in
　　　　　　　　　the personal lives of fictitious characters'

Excluded for similar reasons are CNs in which only one member is taken idiomatically; in the CNs shown in (1.21), for example, the prenominal modifier is used idiomatically while the head noun retains its literal meaning.

(1.21)　　*polka dot, cottage cheese, cathouse, banana republic, handbook, penknife, station wagon, rock music*

Since CNs like those in (1.20)–(1.21) are even less amenable to systematic description than the lexicalized examples cited in (1.16)–(1.19), the former will not be considered further here.

1.4 Terms and Abbreviations

In the following pages, the standard linguistic abbreviations of N, V, Adj, Adv, Prep, NP, S will be used to refer to the node labels (and constituents) of Noun, Verb, Adjective, Adverb, Preposition, Noun

---

[11] Although some people know the historical reasons why the terms *swan song* and *soap opera* are used with their present referents, many others can and do use the terms competently without knowing any relation between the overall meaning and the meaning of the constituent nouns. It is for the latter that these expressions are fully idiomatic, rather than metaphorical. (This is just one way in which individual speakers differ in the way their internal grammar represents particular CN forms; the phenomenon of individual variation in the grammar of CNs is taken up again in Section 7.3.)

Phrase, and Sentence, respectively. It will prove convenient to employ the following abbreviations as well: **CN** for complex nominal, **NC** for nominal compound, **nonpred adj** for nonpredicating adjective, **nom (nonpred) adj** for nominal (nonpredicating) adjective. In addition, $\mathbf{N_1}$ will sometimes be used to refer to the initial member of a two-part complex nominal, whether that member is morphologically a nom adj (as in *atomic bomb*) or a noun (as in *atom bomb*); $\mathbf{N_2}$ will then be used for the head noun.

To refer to the various stages of a derivation, the terms **Logical Structure (LS), Intermediate Structure (IS),** and **Surface Structure (SS)** will be used. Logical Structure refers to the deepest level of a derivation where (in the theoretical framework of generative semantics) all semantic content is organized into labeled trees with just three kinds of nodes: propositions (S), predicates (PRED), and arguments (ARG). The logical PRED nodes will sooner or later become the various derived parts of speech, such as V, N, Adj, and Adv; the logical ARG nodes will become nuclei of various lexical incorporations and insertions, ending up either under derived NP and N nodes or as other categories subsequently derived from them (such as nominal adjectives).

Intermediate Structure refers to the whole sequence of stages, **excluding** the initial and terminal ones (LS and SS), of the derivation of a surface sentence or sentence fragment. (It thus corresponds to Postal's use of "remote structure" in Postal 1970.) Somewhere within this sequence, though not at any independently specifiable level, the predicates and arguments of the LS start to become differentiated and to acquire node labels such as Adj, N, V, or Adv. These labels may or may not survive intact all the way up through Surface Structure, depending on which lexical incorporation processes, if any, are applied before the end of the derivation.

Surface Structure is the final stage of the derivation, representing the syntactic structure of an actual utterance; at this level, the same category label often dominates lexical items whose semantic origins and composition are extremely diverse. (This is notably true of the label Adj which dominates lexical items that may have been derived from antecedent nouns, verbs, adverbs, or various combinations thereof.)

1.5 Order of Presentation

In Chapter 2 evidence is presented to support the first major hypothesis of this work, namely, that a derivation of the adjectives designated earlier as "nonpredicating adjectives" from underlying nouns predicts and explains their allegedly anomalous syntactic and semantic characteristics. Also in this chapter is a comparison of the set of "nominal compounds" widely discussed in the literature with the set of "complex nominals" under analysis here; this comparison will lead to the claim that the "nominal compound" is in most respects a fictitious beast.

Chapter 3 comprises a general statement of the basic principles of which my theory of complex nominals is composed. The question of why languages have complex nominals at all is also raised, and various functional explanations are offered as answers. The chapter concludes with syntactic evidence that CNs function syntactically as nouns, rather than as noun phrases.

The core of the theory follows in Chapters 4 and 5, which are devoted to derivations of complex nominals by predicate deletion and predicate nominalization, respectively. Detailed transformational histories are provided for each of the relevant types of CN, together with justification for the proposed classification. In Chapter 4, the crucial notion of a set of Recoverably Deletable Predicates (RDPs) is introduced in order to explain the specifiable ambiguities evident in those CNs not derived by nominalization, and some implications of this proposal for universal grammar are explored. In Chapter 5, a new analysis of the rules of nominalization in English is presented, together with a discussion of their role in CN formation. Chapter 5 concludes with a section on the issue of recursion and cyclicity in the derivation of all CN types.

A number of ramifications of this theory are then explored in Chapter 6; these include stylistic factors in the grammar of CNs, the implications of the study for lexicographic treatment of CNs, the interaction of semantic and pragmatic principles in our usage of CNs, and finally the contributions of this study to universal grammar. In the penultimate chapter, unresolved problems and putative exceptions are presented, and the fuzzier aspects of this part of the grammar exposed. A summary and review of remaining questions conclude the volume in Chapter 8.

2

NOMINAL ORIGINS OF NONPREDICATING ADJECTIVES

2.1 The Hypothesis of Nominal Origins

The data to be examined in this chapter are illustrated in (2.1) and (2.2).

(2.1)   a. *a rural policeman*
       a chemical engineer
       a corporate lawyer
       a dental appointment
       a linguistic scholar

       b. **a policeman who is rural*
       **an engineer who is chemical*
       **a lawyer who is corporate*
       **an appointment which is dental*
       **a scholar who is linguistic*

(2.2)   a. *a provincial governor*
       a criminal lawyer
       a logical fallacy
       a constitutional amendment
       dramatic criticism

       b. *a governor who is provincial*
       a lawyer who is criminal
       a fallacy which is logical
       an amendment which is constitutional
       criticism which is dramatic

As noted in the previous chapter, the standard transformational analysis of prenominal adjectives in English predicts that the (a) expressions of (2.1) and (2.2) should be derived from intermediate structures which

correspond to the (b) expressions. We note, however, that the forms in
(2.1b) appear to be ill-formed semantically as well as syntactically. We
must therefore seek an explanation for the asymmetry in our grammar
which permits structures like *a policeman who is friendly* to reach the
surface without undergoing Relative Clause Reduction and Adjective
Preposing (which would form *a friendly policeman*) but blocks apparently
parallel structures like those in (2.1b) from expression in SS.

The inadequacy of the standard analysis of prenominal adjectives is
revealed in a different way by the forms in (2.2). Here, the (b) forms **are**
well-formed semantically as well as syntactically. However, the (a) forms
manifest an ambiguity which the (b) forms—from which they are allegedly derived—lack. For example, *a provincial governor* is synonymous
with *a governor who is provincial* only on the reading of 'an unsophisti-
cated governor'. To convey the meaning 'governor of a province', however, we may use only the former and not the latter expression. The
remaining examples of (2.2) show comparable discrepancies.

If we now seek a derivation for the (a) forms in (2.2) when they carry
the **non**shared meanings (i.e., 'governor of a province', 'one who does
criminal law', 'a fallacy in logic', and so forth), we must rule out any
derivation in which the (b) forms would represent an intermediate stage;
since the two sets of forms are nonsynonymous, the theory of generative
semantics requires that we propose distinct underlying structures for each
(while the assumption of meaning-preserving transformations entails that
their nonsynonymity will be preserved throughout the derivation). We
are thus obliged for a second reason to conclude that the "standard"
analysis of prenominal adjectives in English fails to provide a valid derivation for the distinctive set of adjectives shown in the (a) expressions of
(2.1)–(2.2). If, however, a well-motivated alternative derivation can be
elaborated, it not only could serve to correct the overgeneralization
apparent in the earlier theory but also would render moot the question
raised earlier, as to why the forms in (2.1b) never surface; being semantically ill-formed, they would simply never be generated to begin with.

Because these adjectives do not regularly appear in predicate (post-
copula) position,[1] and because they can be shown to be semantically
unlike true predicating adjectives, they are called here "nonpredicating

---

[1] The only cases in which these adjectives do appear in predicate position are those in
which they represent the remnants of entire CNs whose head nouns have been optionally
elided (as in *That interpretation is presidential* from *That interpretation is a presidential
interpretation*). In no case, however, do these adjectives arise alone in predicate position.
For more detailed discussion, see Section 7.2.

adjectives." In this chapter, I will present and defend the hypothesis that these nonpredicating adjectives must be derived from nodes which carry the label N in an immediately antecedent phrase marker. They therefore differ essentially from true, predicating adjectives like *big*, *long*, *stony*, or *joyful* in that members of the latter group must be derived from predicating constituents in IS, and not from any single constituent whose label is N. Because of this hypothesis, and to distinguish these adjectives from those nonpredicating adjectives derived from underlying adverbs (like those in [1.9]), the adjectives illustrated in (2.1) and (2.2) may also be called **nominal nonpredicating adjectives** (**nom nonpred adjs**), or simply **nominal adjectives** (**nom adjs**).

It is important to distinguish between the "nominal adjectives" of this study and the innumerable adjectives of English traditionally called "denominal"; examples of the latter (many more of which can be found catalogued and analyzed in Ljung 1970) include *stony*, *wealthy*, *poisonous*, *joyful*, *disastrous*, *conventional*, *scandalous*, *childish*, *manly*, *artistic*, and *heroic*. All of these are totally unexceptional in syntactic behavior: They can appear in predicate position, may be modified by *very* and other degree adverbials, and in general exhibit all the characteristics of normal adjectives. The essential distinction between the "nominal adjectives" under scrutiny here and the "denominal adjectives" just cited lies in the fact that the former are derived **solely** from nouns, whereas the latter have been derived by the incorporation of a noun with another, predicating constituent.

For example, the "denominal" adjectives *stony* and *wealthy* represent the combination of a nominal element (*stone*[*s*] or *wealth*) with a predicate element (here, the suffix *-y*, with a meaning something like 'having a greater than normal amount of'). The term "denominal" was presumably introduced because the **stem** to which the bound morpheme (such as *-y*, *-ous*, *-al*) is attached can be seen to be virtually or totally identical to an independent noun. This term is misleading, however, in that it seems to imply that the nominal stem, which is morphologically prominent, is also the most important element syntactically or semantically, and that the bound morphemes represented by suffixes like *-y*, *-ful*, or *-ous* serve primarily as markers of changed or derived categories. This is, however, definitely not the case, as may be seen by considering the derivational paths associated with the two sets of adjectives.

"Denominal" adjectives like *stony* or *wealthy* are derived by the incorporation at some point within IS of a constituent labeled N, together with a predicating constituent labeled V or Adj; although the nominal element remains transparently nominal, the predicating element is transformed into the opaque (i.e., less obviously "deverbal") form of a suffix. Morphological transparency is, however, completely irrelevant to semantic interpretation and syntactic category assignment, since in this case it is the more opaque of the two elements (the "deverbal" suffix) which determines both the syntactic fact that the new category label will be Adj and the semantic fact that the new adjective will be a predicating one. If the material now under this Adj label (which has been derived **only in part** from a deeper noun) undergoes no more derivational changes, it will emerge on the surface as a regular, predicating adjective.

In contrast, nominal nonpredicating adjectives derive **all** their semantic content—rather than just part—from antecedent nouns, and it is only at a very late stage in the derivation that the node label Adj is introduced to replace the prior label of N. Since, however, these newly created adjectives have replaced only nouns (and, specifically, nouns which have not become incorporated with any predicating elements), these adjectives emerge on the surface as syntactic and semantic nonpredicators.

With these distinctions in mind, we may now begin the analysis of the sources of nominal nonpredicating adjectives. In Section 2.2, 6 arguments are presented to support the hypothesis of nominal origins for the adjectives in (2.1) and (2.2); the 13 arguments that Postal (1972) provides in support of his proposal for "pseudo-adjectives" are also discussed as relevant supporting evidence. Based on the implications of these arguments, a more precise delineation of the scope of this work is drawn in Section 2.3. The final section of this chapter then takes up the troublesome concept of "nominal compounds," demonstrating that no reliable and mutually consistent tests have yet been proposed that would constitute an acceptable definition of this entity; as a result, the term **complex nominal** as used in this book may not be identified with the term **nominal compound** as it has appeared in previously published literature.

2.2 The Evidence

We turn now to the question of what evidence exists to support the claim that the nonpred adjs in (2.1) and (2.2) are derived from "ancestor constituents"[2] in IS which have the category label of N. The sort of argument we would find most convincing would be an argument showing that these adjectives share syntactic behavior and semantic characteristics peculiar to **nouns,** which would therefore not be manifested by verbs or predicating adjectives. The six arguments I will present, based on six

---

[2] This term is intended in the sense of "derivationally antecedent structures," as used by Postal (1972:1).

distinguishing properties of nouns, are listed here in (2.3).[3] In the paragraphs to follow, each of these arguments will be considered in turn, as well as supplementary evidence drawn from the discussion of pseudo-adjectives in Postal 1972.

(2.3) 1. Nouns may not be immediately preceded by *very*, *quiet*, or other degree adverbials.
2. Nouns conjoin only with other nouns.
3. Nouns may appear after quantifiers, that is, may be counted.[4]
4. Nouns may be categorized by semantic features such as [±definite], [±concrete], [±animate], [±human], and [±common] (or by an equivalent system).
5. Nouns may be analyzed as entering into case relations such as agentive, objective, locative, dative/possessive, and instrumental.
6. Nouns are not subject to the process of nominalization which normally turns predicating elements (verbs and adjectives) into derived lexical nouns.

2.2.1 Nondegreeness

This argument claims that, just as nouns may not normally be preceded by degree adverbials such as *very*, *slightly*, *quite*, or *extremely*, neither may the nonpredicating adjectives we are presently considering. Relevant examples appear in (2.4).

(2.4)

	Nonpredicating modifiers		Predicating modifiers
Adjs:	*very urban riots	Adjs:	very destructive riots
	*very bodily injury		very extensive injury
	*a very electrical conductor		a very efficient conductor
	*very automotive emissions		very sporadic emissions
Ns:	*very (city) riots		
	*very (body) injury		
	*very (electricity) conductor		
	*very (automobile) emissions		

Essentially the same observation was made by Lees (1960:180–181), who suggested that "bona fide" adjectives can be systematically distinguished from "adjectives in compound nominals" (i.e., nom nonpred

---

[3] Arguments 1–5, which first appeared in Levi 1973, are substantially the same as presented in Levi 1975; the sixth is based on a parallel argument provided by Bartning (1976:112–113) for nominal adjs in French.

[4] Note that the existence of mass and proper nouns shows that nounhood is a necessary but not sufficient condition for manifesting this property.

adjs) by the criterion of appearance with *very:* Bona fide adjectives may be preceded by *very*, whereas adjectives in compound nominals may not. To the extent that this generalization is valid, it can serve as both a positive test for true adjectives and a negative test for nom nonpred adjs.[5] Note, however, that my claim that the latter are derived from underlying nouns **predicts** the syntactic irregularity observed—but not explained—by Lees and others. As is shown by the data in (2.4), nom nonpred adjs behave with respect to *very* just like the nouns they come from, and unlike (most) true adjectives.[6]

A theory that derives nom nonpred adjs from underlying nouns can explain not only the data in (2.4), but also some related semantic facts brought out by Bartning (1976:78ff.) in her analysis of nom nonpred adjs in French. Exploiting the three-way distinction (which she bases on Leech 1974:106ff.) between (*a*) **binary oppositions,** which admit no degrees between the two opposites (e.g., *dead/alive*); (*b*) **multiple oppositions,** which generally also lack gradience but which comprise more than two alternatives in a given field (e.g., primary color terms); and (*c*) **polar oppositions,** in which two poles define an entire continuum (e.g., *rich/ poor, old/young*), Bartning observes (1976:79) that in contrast to predicating adjectives, which fall almost entirely into the first and third categories, the vast majority of nominal nonpredicating adjectives represent the **second** category (that of multiple oppositions); moreover, while a very small number of nom nonpred adjs show a binary opposition, there exist **no** instances where these adjectives set up a polar opposition (Bartning 1976:79).[7]

---

[5] It turns out that although this criterion is an accurate test for nom nonpred adjs, it is somewhat of an oversimplification with respect to true adjectives since there exist regularly predicating adjectives that do not happen to be semantically or syntactically intensifiable; *pregnant* is perhaps the paradigmatic example, but one might also cite examples such as *deciduous, immaculate, male, terminal, bankrupt, fatal, unborn, quadrilateral,* and—in its original sense—*unique.* The fact that predicating adjectives include both intensifiable and nonintensifiable examples does not, however, affect the force of the current argument.

[6] Dwight Bolinger has pointed out (personal communication) that "the absence of degree adverbials with nouns . . . does not rule out the existence of degree nouns, e.g., *such a fool* [since] nouns are just as intensifiable as adjectives, i.e., the intensifiable/ nonintensifiable categorization cuts right across the major parts of speech." (This is, in fact, the major thesis of Bolinger's excellent 1972 work, *Degree Words.*) I have argued elsewhere (Levi 1975, Section 2.2.1.1) that the distinction drawn here by Bolinger with respect to surface structure categories is both predicted and explained by a theory such as my own that relates surface syntactic facts on degreeness to a difference in IS category labels (and, ultimately, to a difference in semantic structure itself). Moreover, it remains the case that both "degree nouns" and "nondegree adjectives" are atypical of their respective syntactic categories, a fact which must be reflected in a complete analysis of these forms.

[7] Examples cited by Bartning of multiple oppositions among nom adjs in French may be

The relevance of Bartning's observations to this section lies in the fact that they constitute a reflection in the semantic domain of the same phenomenon that was illustrated in the syntactic data of (2.4), namely, that nom nonpred adjs are not used to denote intensifiable qualities (i.e., qualities which correspond to points along a continuum). Instead, they apparently serve to assign membership to **discrete** subsets of a superset; naturally enough, the number of subsets may be just two (as in binary oppositions like *masculine/feminine*) or some finite number greater than two (as is the case for multiple oppositions, such as that represented by engineering subfields: *chemical/sanitary/biomedical/structural/hydraulic/* . . .). These facts suggest a complementary distribution of function between predicating and nonpredicating adjectives, such that the former are primarily used to assign places along a continuum and the latter to assign membership in specific, discrete subsets of the larger category denoted by the head noun.

The most pertinent aspect of these facts for the present discussion, however, is that both the syntactic and the semantic data follow naturally from a theory which derives nonpredicating adjectives from antecedent nouns; such a theory is able to distinguish these adjectives in a systematic, principled way from regular predicating adjectives and, in so doing, to predict successfully the distinct syntactic and semantic characteristics of each group.

2.2.2 Conjunction of Like Constituents

This argument uses surface syntactic data involving conjunction as a test for more remote syntactic constituency. Although much remains to be learned concerning the constraints which determine well-formed conjunction, it does appear to be true that conjunction is permitted between superficially distinct constituents only if they can be shown on independent grounds to derive from the **same** constituent type at a more remote stage of the derivation. Some of the difficulties yet to be surmounted in

---

seen in (i), while (ii) illustrates the most common of those few pairs whose nom adjs form binary oppositions; note that most have direct parallels in English.

(i) *recherches linguistiques/économiques/sociologiques/etc.*
     développement industriel/économique/culturel/etc.
     régions agricoles/industrielles/minières/laitières/etc.

(ii) *vêtements féminins/masculins*
     administration civile/militaire
     univers matériel/spirituel
     produits naturels/chimiques

constructing such arguments, however, may perhaps be indicated by the variety of surface conjunctions provided in (2.5); these data illustrate just a few of the many complex interactions between syntactic and semantic facts which an adequate theory of conjunction must account for. (Just two of the questions relevant here for which we have not yet found answers are, at what point in the derivation are constituent types to be compared, and what are the criteria according to which two constituent types are to be judged "identical.") Examples (2.5a)–(2.5b) represent grammatical conjunction between superficially unlike constituents, while (2.5c)–(2.5f) represent conjoined structures which are ungrammatical despite the fact that the conjuncts are dominated in SS by the same category label.

(2.5) a. *She left the meeting quietly and with dignity.* (Adv + Prep P)

 b. *Jill is highly respected in her profession and an excellent teacher.* (Adj P + NP)

 c. **She sprayed the bugs on the Ficus with poison and alacrity.* (N + N)

 d. **John vowed to both exonerate and behave himself.* (V + V)

 e. **Lee defended that client initially and brilliantly before Judge Daniels.* (Adv + Adv)

 f. **Luke was a primary and loquacious witness in that trial.* (Adj + Adj)

These data strongly suggest that the grammaticality of conjoined structures must be predicted not on the basis of SS categories but rather on the basis of category membership at a more remote stage of IS (although precisely where in IS remains an open and difficult question). If we adopt this principle as a working hypothesis, at least for present purposes, we observe that my theory makes the prediction that nom nonpred adjs should be (*a*) conjoinable with semantically appropriate nouns and other nom adjs; and (*b*) **not** conjoinable with true adjectives that do not share their nominal origins. The data in (2.6) and (2.7) show that this is indeed the case.

(2.6) Nonpred Adj with N:

 a. *electrical and mining engineers*

 b. *a corporate and divorce lawyer*

 c. *solar and gas heating*

 d. *electrical and water services*

 e. *domestic and farm animals*

  2 NOMINAL ORIGINS OF NONPREDICATING ADJECTIVES

(2.7)   Nonpred Adj only with Nonpred Adj, not with True Adj:

    a. *a civil and* $\begin{Bmatrix} mechanical \\ {}^*rude \end{Bmatrix}$ *engineer*

    b. *anthropological and* $\begin{Bmatrix} ethnographic \\ {}^*respected \end{Bmatrix}$ *journals*

    c. *continental and* $\begin{Bmatrix} oceanic \\ {}^*expensive \end{Bmatrix}$ *studies*

    d. *literary and* $\begin{Bmatrix} musical \\ {}^*bitter \end{Bmatrix}$ *criticism*

Without minimizing the complexities inherent in the analysis of conjunction data in general, we can nonetheless observe that the patterns of acceptability shown here follow directly from a theory which ascribes nominal origins to these nonpredicating adjectives; in contrast, it is difficult indeed to imagine how these data might be predicted, much less explained, in a theory which did not derive these adjectives from antecedent nouns.[8]

2.2.3 Countability

The third argument is based on the fact that nouns (and noun phrases) are the only constituents that may be counted. From this fact, my theory predicts that at least some nonpredicating adjectives should be countable, like the nouns they are derived from and unlike verbs and true adjectives. Since English morphology prevents free quantifier morphemes from modifying surface adjectives, we must look instead at the

---

[8] One recent attempt to explain conjunction data like those in (2.5) is that of Schachter 1977, in which it is claimed that the major constraint on well-formedness of coordinate structures is one which requires identity of both semantic function and **surface** syntactic category for every conjoined constituent. Unfortunately, Schachter's proposal is difficult to evaluate even on its own terms (that is, ignoring for present purposes the difference in theoretical framework between his lexicalist analysis and the generative semantic approach adopted here) because so little detail is provided concerning the crucial notions of "syntactic category" and "semantic function" in terms of which the proposed constraint is formulated. For example, it is impossible to determine whether or not Schachter's theory can predict the ungrammaticality of the starred coordinate structures of (2.7), since we have no way of knowing what range of "semantic functions" his theory would assign to prenominal adjectives like these (i.e., whether or not the same function would be associated with both predicating and nonpredicating modifiers). Nonetheless, it is clear that the well-formed coordinate structures in (2.5a,b) and (2.6) do constitute counterevidence to the proposed constraint, since they demonstrate that grammatical conjunction of nonidentical surface categories is indeed possible.

bound morphemes of quantifying prefixes like *mono-*, *bi-*, *multi-*, and so forth. Consider the data in (2.8):

(2.8)

Prefix + noun	Prefix + nonpred adj	Prefix + pred adj
monoplane	*monochromatic*	**monohigh*
biped	*binational*	**bired*
triangle	*triconsonantal*	**tristrong*
quadrangle	*quadrasonic*	**quadralow*
multicylinder	*multiracial*	**multidense*
polysyllable	*polyphonic*	**polynear*
	omnidirectional	**omnistupid*

These examples indicate that my theory is correct in predicting that numerical prefixes can be attached to both nouns and nom nonpred adjs, but not to true adjectives.

One unexpected result of the prefixation, however, is to remove the composite adjectives from the nonpredicating class, since it turns out that they can appear in predicate position, as shown in (2.9):

(2.9)

a. Simple adjectives	b. Composite adjectives
chromatic analysis	*monochromatic drawings*
**That analysis is chromatic.*	*Those drawings are monochromatic.*
national exports	*binational agreements*
**Those exports are national.*	*Those agreements are binational.*
consonantal structure	*triconsonantal roots*
**That structure is consonantal.*	*Those roots are triconsonantal.*
sonic barrier	*quadrasonic recordings*
**That barrier is sonic.*	*Those recordings are quadrasonic.*
phonic distinctiveness	*polyphonic music*
**That distinctiveness is phonic.*	*That music is polyphonic.*

In view of this syntactic distinction (for which I have no explanation at present), only the Adj + N combinations in (2.9a) may properly be analyzed as CNs, and hence as relevant to our present concerns.[9]

2.2.4 Semantic Classes

The next argument predicts that nom nonpred adjs should be divisible into the same semantic categories that are used to classify nominal

[9] For further discussion of these composite adjectives, see Levi 1975, Section 9.2.

constituents (Ns and NPs) and no others. In presenting this argument, I will use binary features as a convenient shorthand for expressing these semantic categories, but with the explicit stipulation that these features must be considered as nothing more than a particularly expedient notational device, rather than as a theoretical construct of any real utility.

For example, when a noun such as *Boston* or *Markoff* (and/or a derived adjective such as *Bostonian* or *Markovian*) is listed here under [+definite], the feature notation must be understood as actually standing for a whole complex of speaker assumptions about the referents of such nouns (e.g., that there is only one in the speaker's world, or that there is only one which is relevant in that particular conversation, or that the referent is the same as that of a more fully specified description uttered earlier in the discourse). Similarly, the listing of certain nouns as [+animate] and others as [−animate] must be understood as a deliberate oversimplification of some very complex semantic and pragmatic issues. If, however, we keep these qualifications in mind, the binary features which appear in the following paragraphs will serve the useful purpose of simplifying exposition and avoiding unnecessary digressions.

The argumentation in this section focuses on six important semantic divisions among nouns (actually, among full NPs but only single-noun NPs will be considered here); these may be expressed in features as ±definite, ±concrete, ±animate, ±human, +masculine/+feminine, and ±common. All six of these can also be applied to nonpred adjs; the classes thus formed are shown in (2.10).[10]

(2.10)   +definite: *American, Parisian, Markovian, Mexican,* . . .
           −definite: *national, urban, feline, stellar,* . . .

           +concrete: *aquatic, suburban, bodily, lunar,* . . .
           −concrete: *dramatic, constitutional, linguistic, musical,* . . .

           +animate: *senatorial, feline, presidential, Chomskyan,* . . .
           −animate: *rural, electric, acoustic, automotive,* . . .

---

[10] The two features of ±count and ±singular are deliberately omitted from consideration here. The former feature appears irrelevant to adjectival usage, despite the fact that it obviously applies to the nouns from which the nom adjs are derived. We know that *mouth* and *consonant* are [+count], while *country* (in the 'rural' sense) and *electricity* are [−count], but I see no way to apply the distinction sensibly to the adjectives *monthly* and *consonantal* as opposed to *rural* and *electrical*. The ±singular distinction is irrelevant on quite similar grounds. Moreover, since all nom nonpred adjs which are not PPAs (Proper Pseudo-Adjectives; see page 33) are derived from generic source nouns only, their source nouns are presumably unspecified for number. For example, *bovine reproduction* refers to reproduction of either The Cow (generic "singular") or Cows (generic "plural").

+human: *Markovian, presidential, papal, athletic,* . . .
−human: *Bostonian, bovine, ethnographic, consonantal,* . . .

+masculine: *paternal, masculine,* [*Noam*] *Chomskyan,*
       [*Andrew*] *Johnsonian,* . . .
+feminine: *maternal, feminine,* [*Carol*] *Chomskyan,* [*Judy*]
      Levian, . . .

+common: *financial, monthly, urban, musical,* . . .
−common: *Persian, Chomskyan, Elizabethan, Parisian,* . . .

As for nouns, so for nonpred adjs, certain features make others redundant. Thus, in both cases [+human] implies [+animate], and [−common] implies [+definite]. Moreover, the selectional restrictions expressed by these features for nouns carry over quite straightforwardly to the corresponding nonpred adjs. A few examples of parallel restrictions that can be predicted in this way are given in (2.11); the reader can easily supply additional ones. In each case, the grammaticality of the phrases in (b) can be predicted by the grammaticality of the synonymous (and syntactically antecedent) phrases in (a).

(2.11) a. *lies by* $\begin{Bmatrix} presidents \\ {}^*chemicals \end{Bmatrix}$ b. $\begin{Bmatrix} presidential \\ {}^*chemical \end{Bmatrix}$ *lies*

digestion by $\begin{Bmatrix} cows \\ {}^*Paris \end{Bmatrix}$ $\begin{Bmatrix} bovine \\ {}^*Parisian \end{Bmatrix}$ *digestion*

comments by $\begin{Bmatrix} editors \\ {}^*flowers \end{Bmatrix}$ $\begin{Bmatrix} editorial \\ {}^*floral \end{Bmatrix}$ *comments*

intuition of $\begin{Bmatrix} women \\ {}^*Boston \end{Bmatrix}$ $\begin{Bmatrix} feminine \\ {}^*Bostonian \end{Bmatrix}$ *intuition*

glands of $\begin{Bmatrix} mammals \\ {}^*geography \end{Bmatrix}$ $\begin{Bmatrix} mammalian \\ {}^*geographic \end{Bmatrix}$ *glands*

The asterisks in (2.11) mark phrases which would be semantically anomalous in most contexts. The fact that some unusual contexts are conceivable in which such phrases might be used does not invalidate the argument at hand, since it remains true that the acceptability of the phrases in (b) will still be predictable from the acceptability of the phrases in (a). Thus, if a story like *Alice in Wonderland* establishes a context in which flowers can talk, an agentive CN *floral comments* will be perfectly well-formed. But this is predictable precisely from the fact that *comments by flowers* in that context will be equally well-formed.

Note also that some of the starred examples in (2.11b) have acceptable readings which are **not** synonymous with the source phrase suggested in (a). For example, *Bostonian intuition* is a grammatical paraphrase of the expression *intuition that Bostonians have*, since *Bostonians* is a [+human] NP, which is the only sort of NP whose referents would normally be said to possess intuition. In contrast, we would not derive the nonpred adj in *Bostonian intuition* from a strictly locative NP *Boston* (leaving fantasy usages aside once again) since this would violate our assumptions about which entities in the world possess intuition.

2.2.5 Case Relations

One fundamental way of analyzing nominal constituents is in terms of semantically based case relations such as agent, instrument, and location. Since these relations are normally attributable **only** to nouns and noun phrases, my theory predicts that nominal nonpredicating adjectives should also be amenable to just such an analysis. Moreover, just as semantic analyses of NPs in general must assign cases not on the basis of surface configuration (which is relevant only to syntactic case **marking**) but rather on the basis of the underlying propositions from which they are derived, so too the case relations of nonpred adjs must be analyzed in terms of the propositions from which they are derived. Keeping this in mind, we see in (2.12) examples of nonpred adjs expressing agentive, objective, locative, dative, and instrumental case relations within their respective CNs.

(2.12)   Agentive: Objective:
        presidential refusal   *constitutional amendment*
        editorial comment   *cardiac massage*
        revisionist betrayals   *oceanic studies*
        senatorial investigations   *lunar explorations*
        national exports   *dramatic criticism*

<table>
<tr><td>

Locative:
marginal note
marine life
cerebral hemorrhage
urban transit
tropical butterflies

</td><td>

Dative/Possessive:
feminine intuition
feline agility
occupational hazard
judicial prerogatives
planetary mass

</td></tr>
</table>

Instrumental:
manual labor
microscopic analysis
solar generator
aural comprehension
electric calculator

The hypothesis of nominal origins is thus further supported by the fact that it not only correctly predicts that the system of case relations can be usefully extended to include nonpred adjs, but also provides a natural **explanation** for such an extension in terms of the derivational history of these adjectives.

2.2.6 Nominalization

If nonpred adjs are indeed derived from nouns, then we would expect that these adjectives will behave syntactically like nouns in failing to undergo the process of nominalization that regularly applies to verbs and predicating adjectives. This section will show that this is indeed the case.

In her analysis of French nominal adjectives, Bartning (1976:112f.) cites Dell's observation that nominalization applies only to those adjectives which may appear in predicate position (Dell 1970:189ff.); she then exploits this fact to distinguish between the predicating and nonpredicating members of pairs of homophonous adjectives. The relevant data for French, based on Bartning's examples, are shown in (2.13)–(2.15); the (a) examples represent the predicating member of each pair, and the (b) examples the nonpredicating member. (The asterisks in the [b] data are thus to be understood as marking ungrammaticality for the **non**predicating adjective in question.)[11]

---

[11] The lexicographic issue of whether it is more justified to refer to an adjective such as *populaire* in (2.13) as one adjective with two readings, or two distinct but homophonous adjectives, will not be examined here. My general usage, however, reflects the latter option, so that the predicating *populaire* and the nonpredicating *populaire* will be considered here to be two distinct lexical entries.

     2 NOMINAL ORIGINS OF NONPREDICATING ADJECTIVES

(2.13)   a. *une chanteuse populaire*
          La chanteuse est (très) populaire.
          la popularité de la chanteuse

          b. *le front populaire*
          **Le front est (*très) populaire.*
          **la popularité du front*

(2.14)   a. *un garçon ponctuel*
          Ce garçon est (très) ponctuel.
          la ponctualité du garçon

          b. *une source lumineuse ponctuelle*
          **?La source lumineuse est (*très) ponctuelle.*
          **la ponctualité de la source*

(2.15)   a. *l'étudiante nerveuse*
          L'étudiante est (très) nerveuse.
          la nervosité de l'étudiante

          b. *le système nerveux*
          **Le système est (*très) nerveux.*
          **la nervosité du système*

The argument for French carries over straightforwardly into English, as may be seen by the parallel examples in (2.16)–(2.18). Note that it is this distinction between predicating and nonpredicating members of homophonous adjective pairs which makes expressions like *civil engineer, criminal lawyer, diplomatic historian,* and *popular traditions* ambiguous (and hence subject to manipulation in word games and jokes).

(2.16)   a. *mechanical reaction*
          Her reaction was (very) mechanical.
          the mechanicalness of her reaction

          b. *mechanical engineer*
          **The engineer was (*very) mechanical.*
          **the mechanicalness of the engineer*

(2.17)   a. *a nervous applicant*
          The applicant is (very) nervous.
          the nervousness of the applicant

          b. *a nervous disorder*
          **The disorder is (*very) nervous.*
          **the nervousness of the disorder*

(2.18)   a. *a marginal contribution*
        His contribution was marginal.
        the marginality of his contribution

        b. *marginal width [on a page]*
        **The width is marginal.*
        **the marginality of the width*

We must now ask **why** it is that the ability of an adjective to be nominalized is predictable from its ability to appear in predicate position. In attempting to find an answer, we must first note that the nominalization process regularly functions in such a way as to incorporate an abstract head noun with the predicate of an underlying sentence, and to utilize this predicate element as the morphological stem; this is true whether the predicate element is a verb, as in *refusal, decision, transference, growth,* or *knowledge,* or an adjective, as in *mechanicalness, marginality, truth,* or *subtlety.* (Even in those cases where derivational processes build new nouns from old ones, as in *motherhood* or *friendship,* the nouns which serve as stems must be derived from nonreferential nouns in predicate position [i.e., from **predicate** nominals]; these noun stems are thus just as much predicates in their underlying structures as the verbs and adjectives just cited.)

In a sense, we might say there is simply no point (or meaning) in nominalizing an element which is already "nominal." As a consequence, although verbs and **predicating** adjectives (in their role as sentence predicates) are perfect candidates for this transformation, nouns (when **not** serving as predicate nominals) and nominal adjectives are inherently unsuitable since their function in a sentence, and in CNs in particular, is that of a logical argument rather than a predicate.

On another, more systematic, level, we find an answer by examining the meanings which underlie grammatically nominalized adjectives, in order to see what other components in addition to the adjective make up the structure which is the input to the derivational transformation. Without attempting a detailed analysis of adjectival nominalizations, let us simply consider the following data, which show nominalizations of predicating adjectives and some plausible paraphrases for them.

(2.19)   a. *I was impressed by* $\left\{ \begin{array}{l} \textit{her civility} \\ \textit{the fact that she was civil} \end{array} \right\}$

        under the circumstances.

        b. $\left\{ \begin{array}{l} \textit{His nervousness} \\ \textit{The fact that he was nervous} \end{array} \right\}$ *hurt him in the interview.*

c. *They took pains to insure* $\left\{ \begin{array}{l} \textit{the legality of all procedures} \\ \textit{that all procedures be legal} \end{array} \right\}$

(2.20)   a. $\left\{ \begin{array}{l} \textit{Her nervousness} \\ \textit{The degree to which she was nervous} \end{array} \right\}$ *was worse than usual.*

b. *She tried to reduce*
$$\left\{ \begin{array}{l} \textit{the mechanicalness of her acting} \\ \textit{the degree to which her acting was mechanical} \end{array} \right\}$$

c. $\left\{ \begin{array}{l} \textit{Sam's civility} \\ \textit{The degree to which Sam was civil} \end{array} \right\}$
was less than overwhelming.

Although these adjectives are potentially ambiguous along the predicating–nonpredicating lines illustrated earlier in (2.16)–2.18), it is only the predicating adjectives which are illustrated here; this is demonstrated by the fact that the adjectives in (2.19)–(2.20) undergo nominalization just like normal predicating adjectives (cf. *tall, tallness; loyal, loyalty; elegant, elegance*) and unlike nonpredicating adjectives (cf. *rural, *rurality; lunar, *lunarity; cardiac, *cardiacity*).

The paraphrases in (2.19)–(2.20) suggest that the structures which underlie grammatical nominalizations like those in the (a) examples of (2.16)–(2.18) regularly include, in addition to a head noun like FACT or DEGREE, a complement S in which the adjective is in predicate position.[12] Since, however, nonpredicating adjectives by definition are blocked from occurring in predicate position, they will never appear in the appropriate input structures for this nominalizing transformation.

The argument up to this point has shown that the ability of an adjective to be nominalized is predictable from its ability to appear in predicate position. What remains to be demonstrated is the relevance of this correlation to the hypothesis of nominal origins for nonpredicating adjectives. In order to make this clear, we must show that such a hypothesis predicts that the derivational history of such adjectives excludes a stage in which they appear in predicate position; if this is true, the failure of these adjectives to undergo nominalization will follow automatically.

[12] I am not in a position to assert categorically that the nominalizations in (2.19)–(2.20) **must** come from the same source structures as the paraphrases that accompany them; I am therefore using the paraphrases only to suggest what these structures might be. A more thorough analysis remains to be carried out, presumably along the lines of that provided in Chapter 5 for the nominalization of verbs.

The standard transformational derivation of prenominal predicating adjectives from relative clause constructions in which these adjectives appear in predicate position will be considered here to be well-motivated; thus, *a nervous applicant* is appropriately derived from the same structure that may also surface as *an applicant who is nervous* (details of tense aside). On the other hand, a comparable derivation for **nouns** which occur as prenominal modifiers (i.e., as the first element in CNs) is not at all motivated, as can be seen from the data in (2.21); the presumed source structures (shown on the right) are not merely nonsynonymous with the CNs to their left, but the putative relative clauses (with these particular nouns in predicate position) are in fact syntactically ill-formed and semantically incoherent. As a result, the forms on the right could not serve as underlying structures for **any** well-formed nominals in English.

(2.21) a. *systems engineer* ≠ **engineer who is systems*
        b. *stomach disorder* ≠ **disorder which is stomach*
        c. *foundation offices* ≠ **offices which are foundation*
        d. *page width*   ≠ **width which is page(s)*

It thus appears that although nouns do (at times) share the role of prenominal modifier with predicating adjectives, the two constituent types must follow quite distinct derivational paths in order to reach prenominal position.

Keeping this distinction in mind, we note that the hypothesis of nominal origins for nonpredicating adjectives necessarily predicts that the derivational history of these adjectives must be like that of **nouns** used as prenominal modifiers (since it is precisely those nouns which I claim are the antecedents for those adjectives) and unlike that of normal adjectives. The data of (2.22) provide at least partial support for this prediction by showing that the relative-clause source appropriate for predicating adjectives is as unacceptable for nom nonpred adjs like those in (2.22) as we have seen it to be for nouns like those in (2.21).

(2.22) a. *naval engineer*   ≠ **engineer who is naval*
        b. *a nervous disorder* ≠ **a disorder which is nervous*
        c. *legal offices*   ≠ **offices which are legal*
        d. *marginal width*   ≠ **width which is marginal*

It thus appears that there is as little motivation for deriving nom nonpred adjs via an intermediate stage in which they appear as predicate elements in relative clauses as there was for so deriving nouns. Note, however, that if these adjectives do **not** pass through a stage in which they appear in predicate position, they will never meet the structural description for the nominalization transformation at issue here (namely, one which incorporates a **predicate** element with an appropriate abstract head noun such as FACT, DEGREE, or perhaps STATE). These facts therefore provide additional support for the hypothesis of nominal origins, since the failure of nonpredicating adjectives to undergo the nominalization transformation which regularly applies to predicating adjectives follows in a principled fashion from a theory in which the former are derived from antecedent nouns.

The data in (2.21) and (2.22) demonstrate that the derivational histories of nouns and nonpred adjs acting as prenominal modifiers are alike according to a **negative** criterion, namely, that they cannot have been derived in the same manner as regular, predicating adjectives. To complete the argument, however, we must ascertain just what the distinctive derivations of these forms must be, and so demonstrate in a positive fashion that both the nouns and adjectives appearing as prenominal modifiers in CNs have derivations that are alike in all but the most superficial detail. This complex task will be accomplished in the chapters that follow.

2.2.7 Evidence from Postal 1972

Paul M. Postal is one of the few linguists (transformational or otherwise) to have studied nonpredicating adjectives as a distinctive set. His research (Postal 1969, 1972) has, however, been more limited in scope than mine in that it has focused almost exclusively on the set of "Proper Pseudo-Adjectives" (PPAs) first reported on in Postal 1969, that is, on those "pseudo-adjectives" which must be derived from proper nouns, as in *American refusal* or *Persian application*. (In this respect, his studies complement the present work which has in general been limited to CNs whose first element is a common noun or its adjectival equivalent.)

It seems clear that Postal uses the term "pseudo-adjective" to denote just those forms which I have termed "nominal nonpredicating adjectives."[13] As a convenient distinction, however, the term "pseudo-

---

[13] As indicated earlier, the term "nonpredicating adjective" can appropriately be applied both to those adjectives that I claim are derived from nouns and to those adjectives presumably derived from adverbs (as in *potential enemy, former husband, frequent visitors*) which have been excluded from treatment here; this is because neither of these sets can be derived via the standard predicate-to-attributive-position shift. For similar reasons, one could use the term "pseudo-adjective" to cover both groups; it appears, however, that "adverbial pseudo-adjectives" were not considered by Postal. See the diagram in Levi 1973:339 for a delineation of the various sets of data in question.

adjective" (or PA) will be retained to refer to the examples considered in Postal 1969 and 1972, while "nominal nonpred adjs" will continue to indicate the broader set of data examined in this study.

Postal's major hypothesis concerning PAs is that they must be transformationally derived from underlying nominals; he writes specifically (Postal 1972) that:

> Each Surface Structure containing a pseudo-adjective is the final element of a *Derivation* in which there is an earlier or more remote structure containing an NP constituent *corresponding* [in the sense of corresponding nodes in global grammar, as described by G. Lakoff elsewhere] to the Surface Structure pseudo-adjective. That is to say, roughly, that pseudo-adjectives have NP ancestor constituents in the way preposed adjectives in general have full restrictive relative clauses . . . [p. 1].

The arguments that he then presents in support of his claim involve primarily anaphoric relations and constraints on NPs, with respect to all of which he shows that pseudo-adjectives behave just like the NPs from which he claims they are derived.[14] His 13 arguments involve the Coordinate Structure and Command Constraints, as well as such transformations as Equi-NP Deletion, Super-Equi-NP Deletion, Reflexivization, Symmetrical Pronoun Elision, and anaphoric transformations introducing lexical items such as *that*, *the same*, *the former* and *the latter*, and *own* (Postal 1972, Part II).

The point of each of his arguments is that an analysis of PAs not based on a denominal derivation would have to state these 13 principles twice—once for all NPs, and then once again, with considerable loss of generality, for pseudo-adjectives. This is, of course, the same point that underlies the 6 arguments presented in the preceding sections. A theory that does not derive nom nonpred adjs from nouns could not predict these 6 major aspects of syntactic and semantic behavior which are shared by nouns and nom nonpred adjs. Moreover, such a theory would still be confronted by the problem of characterizing this set of anomalous adjectives in order to distinguish them from true adjectives.

The first part of Postal 1972 contains 13 specific arguments concerning deletion, anaphora, and transformational constraints, the general import of which has just been indicated. In the course of providing a rebuttal to Chomsky 1972a in Part II, however, Postal presents a number of addi-

---

[14] Actually, he claims they behave **almost** exactly like their NP ancestors. The one difference he finds is that PAs cannot serve as antecedents for outbound anaphora in the same way that their ancestor NPs can (Postal 1972:9). This distinction, which I believe is a consequence of the fact that the ancestor NPs become part of a new N constituent created by the process of CN formation, is discussed in more detail in Levi 1977:330–333.

   2 NOMINAL ORIGINS OF NONPREDICATING ADJECTIVES

tional arguments that deserve further attention here; these arguments all pertain to case relations, and provide further support for the hypothesis of nominal origins for pseudo-adjectives.

Postal observes in 1969 and again in 1972 that his thesis is supported by the fact that NPs and their corresponding PAs occur in **complementary distribution** on the surface. Thus he argues (Postal 1969):

If PPA do not have a nominal derivation, a special statement of syntactic restrictions will be required to handle facts like those in:

(77) a. the invasion of America by France
      b. France's invasion of America
      c. the French invasion of America
      d. *the French invasion of America by Portugal
      e. *France's French invasion of America

That is, the'PPA *French* in such cases is in complementary distribution with the genitive NP/*by* phrase NP, that is, with the agent NP. Under the proposal of nominal derivation, this is an automatic consequence [p. 220].

As Postal points out, the relationship of complementary distribution observed here between PPAs and certain syntactic constructions (genitives, "*by* phrases," and so forth) can not only predict a wide variety of ungrammatical structures (such as those in [77d] and [e] above), but can also disambiguate ambiguous occurrences of PPAs, and in general account for a number of otherwise unexplainable facts of semantic interpretation.

To illustrate disambiguation, consider the data in (2.23), adapted from Postal 1972:34.

(2.23)   a. *the Martian expedition*
          b. *the expedition to Mars/on Mars/by Martians*
          c. *the Martian expedition by America*
          d. *the American Martian expedition*
          e. *the Martian expedition to Venus*
          f. *the Martian Venusian expedition*

(2.23a) is ambiguous among (at least) the directional, locative, and agentive readings spelled out in (2.23b). However, the addition of an agentive phrase *by America* in (2.23c) rules out an agentive reading of *Martian* for that example (on the grounds that only one NP can be agent at a time),[15] and we then are left with just the directional and locative possibilities (i.e., 'expedition to Mars' or 'expedition on Mars').

---

[15] Coordinate NPs could be agents, of course, but see the following discussion of (2.25) for why this reading must be ruled out for the pseudo-adjective constructions shown here.

The same principle of complementary distribution compels us to attribute different case relations to the two PPAs in (2.23d); one will be agentive, and one will be locative or directional. Similarly, the directional *to Venus* in (2.23e) disambiguates the prenominal PPA, so that we get an agentive reading for *Martian*, which is retained in our reading of (2.23f).[16]

Our ability to understand these data, and to predict many more along the same lines, is clearly dependent on the generalizations expressed here in terms of semantic case relations. Note, however, that it is only the claim that these PPAs are derived from NPs that permits us to profit from an analysis based on case relations since it is only NPs, and not regular adjectives, to which case relations may be meaningfully attributed.

We can easily extend Postal's data on PPAs to nonpred adjs derived from common nouns; such an extension may be seen in (2.24).

(2.24) a. Agentive: *a judicial attack*   { *on bureaucrats* / **by bureaucrats* / **by judges* }

        b. Objective: *literary criticism*   { *by professors* / **of music* / **of literature* }

        c. Locative: *an aerial attack*   { *on the missile sites* / **under the ground* / **from the air* }

        d. Instrumental: *aural comprehension*   { *of Swahili* / **using the eyes* / **using the ears* }

        e. Dative/Possessive: *presidential power*   { *over appropriations* / **of senators* / **of presidents* }

These data further illustrate the wide range of predictions based on case relations which the nominal origins hypothesis permits us to make. Note that for each CN in (2.24), there are two starred phrases given; the first is ruled out as repeating the **case** expressed by the prenominal adjective (as in the double instrumental phrase **aural comprehension using the eyes*), while the second is ruled out as repeating the nominal ancestor of that adjective as well as the case that **it** expresses (as in **literary criticism of literature* on a double objective reading). On the other hand, the fully acceptable phrases (such as *a judicial attack on*

<hr>

[16] No attempt has been made here to specify the principles according to which a sequence of nom adjs expressing different case relations must be ordered within an NP (as in *American Martian expedition* or *suburban racial integration*). For some preliminary proposals on this subject, see Levi 1975, Chapter 8: "Multiple Nonpredicating Adjectives."

bureaucrats) are neither redundant nor contradictory with respect to the case relations expressed by the prenominal and postnominal modifiers.

Postal also points out examples of ungrammatical sentences that can be explained—in a denominal theory of pseudo-adjectives—by the Coordinate Structure Constraint (or CSC; see Ross 1967). The relevant data are in (2.25):

(2.25) a. *the attack on Spain by Persia and France*
  b. **France's attack on Spain by Persia*
  c. **the French attack on Spain by Persia*

We have already seen that (2.25c) would be ruled ungrammatical on the basis of case conflict, since an agentive reading of *French* would conflict with the agentive reading of *by Persia*. But why do we not get a conjoined agentive reading, equivalent to (2.25a)?

In a denominal theory, a coordinate reading of either (2.25b) or (2.25c) would have to be derived from the same structure underlying (2.25a). But the CSC blocks the movement of a single conjunct out of a coordinate structure, and hence would block the derivation of either (2.25b) or (2.25c) from (2.25a).[17] In contrast, a theory that did not derive these adjectives from nominal ancestors would have to strangely extend, or otherwise duplicate, the CSC in order to account for such data.

In a variety of ways, then, Postal's work on Proper Pseudo-Adjectives provides detailed and well-documented arguments which corroborate the major claim of this chapter concerning the nominal origins of nonpredicating adjectives.[18]

2.3 Scope Extension

My research on complex nominals had its origin in the problems posed by nonpredicating adjectives. The first stage in this work culminated in the hypothesis to which this chapter is devoted, namely, that these adjectives must be derived from underlying nominal nodes. The argumentation in Section 2.2 has, I trust, provided ample support for such a claim.

[17] Interestingly enough, there is a set of PPAs (brought to my attention in this context by Jim McCawley, and also noted in Postal 1972:69) that are derived directly from such a coordinate structure, as in *Anglo-American cooperation*, *Sino-Soviet relations*, or *Indo-Iranian languages*. Note, however, that these obey the CSC, in that the entire coordinate structure has been preposed, rather than just a single conjunct.

[18] Additional comments on this regrettably unpublished study by Postal may be found in Levi 1975, Sections 2.2.2 and 6.1.3.

The logical consequence of this development proved to be that only a unitary explanation could account for the derivations of the syntactically and semantically parallel pairs shown in (2.26):

(2.26)

a. Nonpred adj + noun	b. Noun + noun
hydraulic engineer	*systems engineer*
acoustic research	*motivation research*
criminal lawyer	*divorce lawyer*
digestive system	*communications system*
electrical service	*water service*
presidential power	*student power*

That is, it must be the case that the derivations of the examples in (2.26a) parallel those of the corresponding expressions in (2.26b) all the way through IS up to a very shallow level; then, in the case of the (a) examples, the noun in prenominal position is transformed by a late, optional rule into its corresponding nom nonpred adj. In this way, *sound research* is transformed into *acoustic research*, or *digestion system* into *digestive system*. However, since the application of this late rule effects only the most superficial of changes, the semantic interpretation and syntactic behavior of CNs like those in the two columns of (2.26) will remain the same, regardless of whether this rule has applied. It is for this reason, in fact, that we find in English pairs of CNs like those in (2.27) which are totally synonymous, despite the fact that the prenominal modifiers are adjectives in one case and nouns in the other:

(2.27)   a. *atom bomb*   b. *atomic bomb*
            mother role   *maternal role*
            industry output   *industrial output*
            ocean life   *marine life*
            language skills   *linguistic skills*
            city parks   *urban parks*

The scope of the present study, then, necessarily encompasses not only the nonpredicating adjectives that were my original research focus but, in fact, all the expressions exemplified by the data in (2.26)–(2.27). To describe this larger range of examples, the term **complex nominal** has been introduced and will be used throughout this work. It replaces the term **nonpredicate NP** which was (erroneously) used in Levi 1974 and 1975 but which must be rejected on the grounds that these expressions are not NPs at all, but rather single nouns (a claim I will document in Section 3.4).

The term *complex nominal* thus refers to that syntactic construction dominated by an N node and composed (in its simplest form) of a head noun preceded by a modifier which is either another noun or a nominal adjective. The term **nominal** is chosen here rather than **noun** since I suspect that most linguists, like other people, automatically think "single orthographic unit" when they hear the latter term, and are in no way accustomed to using it either for very long sequences or for forms containing superficial adjectives; using the more neutral term *nominal* to include all the forms at issue here thus sets up no conflict with either our preprofessional instincts or, for that matter, our previous professional readings. Keeping "nominals" and "nouns" separate also permits us to preserve the distinction between lexical items formed by the recursive rules of CN formation and those that are not (i.e., between CNs and the "simple" nouns of which they are composed).

On similarly pragmatic grounds, the modifier **complex** is chosen over **compound** because the latter is too closely associated in linguistic tradition with those forms composed of just two nouns, and typically exhibiting fronted stress, such as *stéam boat* and *car thief*; its scope is therefore too narrow for present purposes. (Moreover, as the concluding section of this chapter will demonstrate, the terms traditionally used to denote such nominals, namely, *compound noun* and *nominal compound*, have never been defined in any rigorous or consistent fashion, thus making the use of either term even less attractive.)

In conclusion, I would emphasize that the broadening of my research focus from the subject of nonpredicating adjectives to the more extensive domain of complex nominals was a direct result of investigating the nominal origins hypothesis that has been defended in this chapter. Because this hypothesis led inexorably—albeit somewhat unexpectedly—into two important areas of linguistic study that had seldom been linked to these particular adjectives, namely, "compound noun" formation and nominalizations (to be analyzed in detail in Chapters 4 and 5, respectively), the extension of scope that was forced by the adoption of this hypothesis permitted the integration of three rather disparate topics into the unified analysis presented in this book. The nominal origins hypothesis is thus supported not only by the numerous arguments cited earlier, but also by the analytic fruitfulness and explanatory power which it introduces into this part of our linguistic theory.

2.4 Nominal Compounds versus Complex Nominals

In broadening the scope of my inquiry from just those CNs which contain nonpredicating adjectives to the larger set which includes those composed of two or more nouns, I was led as a matter of course to consider the very extensive literature already published on the subject of **nominal compounds** or **compound nouns.**[19] Since the reader familiar with some part of this literature may well be wondering why I have rejected these more familiar terms in favor of the relatively novel term **complex nominal,** it would be appropriate at this point to examine the reasons for adapting this terminological distinction.

We may note to begin with that no linguist studying complex nominals could afford to ignore previously published work on nominal compounds (or NCs) for the simple reason that the two sets of data in question show very extensive overlapping. When, however, I discovered that the two sets are nonetheless not entirely coextensive (the inclusion or exclusion of forms with adjectival modifiers being just one difference between the two sets), it became essential for me to answer these two crucial questions: (*a*) Do NCs constitute a **definable** subset of all CNs, and (*b*) if so, what are the precise criteria according to which NCs may be isolated from other, noncompound CNs? Without such an understanding of the intended domains of different analyses, I could make no meaningful comparison between my own theory and any of the numerous works that had already appeared on NCs.

Unfortunately, despite the extensive and intensive attention devoted to nominal compounds in the linguistic literature of the last 40 years or so, very few authors have included in their discussions any attempts to provide a rigorous and formal definition of the object of their studies. In fact, a number of authors are quite explicit in eschewing definitions, contenting themselves instead in treating compounds as some sort of intuitively obvious, "pretheoretical" construct whose study may be profitably pursued even in the absence of a "watertight definition." (See, for example, Botha 1968:153 and Lees 1960:119, 180–181.) Others, including Jespersen (1942), at least attempt to spell out clear and consistent criteria for identifying compounds but eventually conclude that the task is an apparently hopeless one.

In searching for a reliable definition of NCs, many authors have turned with varying degrees of confidence to three purportedly valid

[19] In the area of English nominal compounds, this literature includes the comprehensive studies by Koziol (1937), Jespersen (1942), Lees (1960), Marchand (1969), and Brekle (1970); in addition, Hatcher (1960), Bolinger (1967), Gleitman and Gleitman (1970), and Newmeyer (1975) provide more restricted studies in this area. For studies of nominal compounds in a number of other languages, the reader may wish to consult Barbaud 1971 and Bartning 1976 on French (the former on overtly N–N structures, the latter on nom nonpred adjs); Botha 1968 on Afrikaans; Dede 1977 and Lees 1960 on Turkish; Levi 1976 and Reif 1968 on Modern Hebrew; Li 1971 on Chinese; and Whitney 1879 on Sanskrit.

criteria for distinguishing nominal compounds from other "nominal phrases"; these criteria are the tests of **fronted stress, permanent aspect, and semantic specialization.** Let us examine each of these in turn in order to see why these criteria do not successfully define the "nominal compound" and hence why they have not been included here as part of the defining characteristics of CNs.

Fronted stress, to begin with, is the phenomenon which allows us to distinguish the "compound" form *bláckbìrd* from the attributive-adjective-plus-noun phrase *blâck bírd;* similarly, fronted stress may disambiguate pairs such as *Frénch teàcher* 'one who teaches French' and *Frênch teácher* 'teacher who is French', or *báby photògrapher* 'one who photographs babies' and *bâby photógrapher* 'photographer who is a baby'.

A systematic inspection of the data in question reveals, however, that the test of fronted stress is of limited value at best.[20] That is, the presence of fronted stress seems to correlate positively with (alleged) compound status, while the absence of fronted stress fails to correlate in any systematic way at all with such status.[21] Thus, although (2.28) contains a representative sampling of forms repeatedly cited as compounds, only those in (2.28a) exhibit fronted stress. On the other hand, the forms in (2.28b) resemble those in (2.29) in having normal (nonfronted) stress patterns, but the former are usually considered to be compounds whereas the latter are not. (Examples of CNs with nonpred adjs are included to show that they parallel the N–N compounds in this particular phonological aspect, as well as in their syntax and semantics.)

(2.28)   (Alleged) compounds

   a. **Fronted stress**
      N–N: *apple cake, battle fatigue, stock exchange, moth hole, girl friend, courtyard, garden party, blood test, money order*

      Adj–N: *floral wreath, vocal range, tidal wave, saline solution, nervous system, polar bear, postal service, athletic club*

---

[20] Accordingly, Jespersen (1942:135–136) explicitly rejects it, and Lees (1960:119f., 180ff.) adopts it more for convenience in limiting the scope of his study than for any predictive value.

[21] In speaking of (alleged) compound status here and throughout this section, I mean, roughly, that yet-to-be-formally-defined quality that permits surprisingly general agreement on at least the clearest candidates for compoundhood (e.g., *doghouse, apple cake, atom bomb*). As the discussion will make clear, however, no more precise definition is available from any source that is of use in less clear cases.

b. **"Normal" stress**

N–N: *apple pie, child prodigy, class reunion, string quartet, stone wall, government property, voice vote, surface tension*

Adj–N: *electric shock, atomic bomb, industrial revolution, electrical engineer, polar expedition, feminine intuition*

(2.29)   (Alleged) noncompounds
N–N: *student power, family antiques, mosaic floor, city parks, lake temperature, committee decision, manager efforts, concussion force*

Adj–N: *industrial area, maternal wrath, feline agility, judicial discretion, suburban crabgrass, parental visitors, spinal inflammation*

Three of the serious problems that arise in analyzing these data are (*a*) the existence of CN pairs whose members are semantically parallel or even synonymous but in which stress assignment varies, as in *apple píe* and *ápple cake*, or *fáther figure* and *parental fígure*; (*b*) the fact that one's judgment of stress assignment can be affected by unconscious imposition of contrastive stress on the prenominal modifier; and (*c*) the existence of dialectal differences in stress assignment for some forms, such as *chicken soup, cardiac arrest*, or *picture window*.

In light of all these difficulties and inconsistencies, we can only view as amply justified the conclusion drawn by Jespersen, Lees, and others that fronted stress cannot be considered to be a defining characteristic of nominal compounds in English. (Whether the distribution of fronted stress in these forms is entirely arbitrary, or the result of factors as yet undiscovered, remains an open question.)

Let us consider next the possibility of using **permanent aspect** as a test for true compounds. Most writers on compounds have noted that the bond between elements of a compound seems to reflect a more **permanent,** or at least **habitual,** bond than is the case in noncompound constructions. Gleitman and Gleitman (1970), for example, write:

> Both syntactically and semantically the compounding process reflects *a more intimate and irreversible relation* among its elements than do nominalizing transformations. . . . Semantically, there seems to be a "permanent" association between the elements of a compound, rather than the momentary, one-time-only, relation that may be implied for the elements of a phrase [p. 72; emphasis added].

Thus, the compound *water bugs* (or the equivalent *aquatic insects*) could only describe insects that have some permanent association with water,

                        2 NOMINAL ORIGINS OF NONPREDICATING ADJECTIVES

such as living in or around it, rather than some accidental and ephemeral connection, such as falling into it. Similarly, a term such as *meetinghouse* is used appropriately only for a house in which meetings are regularly held (and/or for which meetings are its primary purpose), but not to describe a house in which one happened to meet some friends on one occasion.

Unfortunately, permanent aspect turns out to be no more successful than fronted stress in predicting membership in the set of nominal compounds. Numerous examples of compound forms (i.e., forms cited by various authors as compounds, presumably on the grounds that they pass one or both of the other alleged tests for compoundhood, namely, fronted stress and semantic specialization) can be listed which cannot be analyzed as representing some "permanent" or "habitual" relationship between the two surface elements, as shown by the data in (2.30).

(2.30)　　N–N:　*moth hole, virus infection, drug deaths, heart attack, weight loss, car crash, diamond heist, birth trauma, vapor lock*

　　　　　　Adj–N:　*tidal wave, viral pneumonia, clerical error, spinal tap, axial stress, electric shock, traumatic arthritis*

The data in (2.30) represent one type of potential counterexample to the claim that permanent aspect may be used to define compounds; a second type would be comparable structures which are judged to be noncompounds but which do exhibit permanent aspect. When we look for instances of the latter, however, we discover that they are more readily found among CNs with nonpred adjs, and that parallel examples with a N–N structure are as often as not intuitively judged to be compounds. Consider the data in (2.31). (Where a form may be interpreted aspectually in several ways, only a reading in which the association is a permanent one is intended here.)

(2.31)　　a. *American villages*　　　　b. *mountain villages*
　　　　　　African rivers　　　　　　　*state rivers*
　　　　　　feline agility　　　　　　　*acrobat agility*
　　　　　　occupational hazards　　　*job hazards*
　　　　　　pedal extremities　　　　　*incisor teeth*
　　　　　　infantile egocentricity　　*infant egocentricity*

　　　　　　polar temperatures　　　　c. *jungle temperatures*
　　　　　　bovine stupidity　　　　　　*management stupidity*
　　　　　　atheist attitudes　　　　　　*bureaucrat attitudes*
　　　　　　popular indifference　　　　*student indifference*
　　　　　　judicial prerogatives　　　*faculty prerogatives*

The examples in (2.31a) comprise CNs which I judged to contain that semantic element of permanent association in question here; the examples in (2.31b) and (2.31c) comprise N–N parallels, of which the former seem to be compounds (which here means only that they strike me as being somehow like the examples typically cited in the literature on NCs) while the latter do not. The fact that the syntactic category of the prenominal element influences one's judgment of compoundhood suggests that the combination of an N–N structure with permanent aspect correlates strongly (though not invariably) with alleged compound status. Nonetheless, the existence of the data in (2.30), (2.31a), and (2.31c) indicates that we cannot rely on permanent aspect as a defining characteristic of compounds; at best, it must be viewed as being positively correlated with judgments of compoundhood in much the same way as is fronted stress. (It is significant, however, that the presence of permanent aspect in CNs with nonpred adjs produces a judgment of compoundhood much less frequently than in the case of N–N CNs, since this discrepancy strongly suggests that such judgments have little to do with the systematic syntactic and semantic parallels that may be demonstrated between N–N and Adj–N CNs and much more to do with the most obvious and superficial feature of surface morphological form.)

We turn now to the third possible criterion, that of semantic specialization. This criterion is equivalent to the claim that all true compounds have become at least partly lexicalized, so that their full semantic specification must be learned on an individual basis, rather than being "recoverable" from (*a*) the meaning of the two surface nouns, plus (*b*) whatever general relations are assumed to be **regularly** expressed (whether overtly or not) by compound forms. (That it is the combination of both these sources of meaning that must be specified as insufficient stems from the fact that **no** linguist analyzing compounds would ever claim that the surface elements alone suffice for semantic interpretation; indeed, it is precisely this tantalizing absence of an overtly expressed relation that has inspired so much discussion of the semantic structure of compounds.)

Despite the fact that this claim in effect denies the undeniable (i.e., suggests that speakers and listeners cannot make use of spontaneous and creative coining of nominal compounds without a breakdown in communication), such a suggestion is made with surprising frequency, at least in the traditional literature. Thus Allerton (1972) criticizes Brekle (1970) for giving no clear-cut definition of compound words, further asserting that, "This specialization in meaning [is] in my view one of the central characteristics of compounds [p. 332]." Similarly, Marchand (1969:25) characterizes this difference as one of a "permanent lexical relation" in

        2 NOMINAL ORIGINS OF NONPREDICATING ADJECTIVES

true compounds, as opposed to a "mere syntactic relation" in noncompounds.

The CN *battle fatigue* may be used here to illustrate this claim. If this expression were "merely" a nominal phrase and not a true compound, so the argument goes, the full meaning of this phrase would be expressible in terms of the two surface nouns plus one of the general semantic relations attributable in a given theory to nonlexicalized CNs; in the present theory, as we shall see, such a form could be generated by the deletion of the predicate CAUSE from an intermediate structure comparable to *battle-caused fatigue*. And yet we know that this expression does not (in common usage) refer to a feeling of simple tiredness experienced after just any kind of battle, but rather to a condition of psychic trauma induced by a soldier's direct experience of the horrors of war on the battlefield. It is this additional semantic specialization that allegedly determines that *battle fatigue* should be classed as a compound while *housecleaning fatigue*, for example, should not.

It should come as no surprise to the reader to learn that this third criterion fails as badly as the others in constituting a definitive test for compoundhood, even though it too seems to have some limited use as yet another positively correlated feature. There is, however, an obvious functional explanation for this particular correlation. That is, the fact that CNs which do exhibit semantic specialization always fall into the intuitively defined class of compounds is actually a reflection of (*a*) the fact that CNs which are used frequently enough as names for familiar objects or entities are often treated (i.e., learned and/or stored) as nonderived, unitary lexical items, which are then subject to just the same accretions of semantic specialization as any monomorphemic entry in our lexicon; and (*b*) the fact that these highly familiar forms, as opposed to newer or more ephemeral ones, predominate among citations in almost all compounding studies (with the exception of analyses devoted specifically to novel compounds, such as Zimmer 1971, 1972; and Downing 1975).

Although the patently untenable claim discussed earlier (to the effect that compounds—or CNs—are not subject to creative use) would oblige us in any case to reject this third criterion, an additional reason for judging it to be useless in isolating compounds may be found in the fact that there are innumerable "compound" forms (along with the expected noncompounds) which show no signs of being lexicalized at all. This may be seen by comparing the allegedly paradigmatic set of semantically specialized compounds shown in (2.32) with the counterexamples in (2.33), all of which represent compounds that lack any significant degree of semantic specialization.

(2.32)   N–N: *battle fatigue, milieu therapy, queen bee, future shock, gunboat, girl friend, windmill, bedbug, heart attack*

Adj–N: *polar bear, tidal wave, civil engineering, clerical error, manual labor, nervous system, solar system*

(2.33)   N–N: *home life, salt water, lemon peel, sugar cube, auto mechanic, blood test, disease germ, moth hole, air pressure*

Adj–N: *marginal note, urban transportation, axial stress, national exports, avian sanctuary, musical criticism*

It is particularly instructive to note that among the CNs which pass this particular test for compoundhood may be found forms which fail the previous two; thus, *heart attack, queen bee,* and *milieu therapy* all have specialized meanings but the first does not incorporate permanent aspect, and the latter two lack fronted stress. Similarly, *lemon peel* and *auto mechanic* flunk the semantic specialization test while passing both the fronted stress and the permanent aspect tests. Thus, whatever encouragement one may derive from identifying three features whose presence, at least, appears to have some predictive value in identifying compounds must surely be offset by the demonstrable fact that these three tests are mutually inconsistent.

May we conclude from the preceding discussion that there are **no** clear and consistent criteria according to which an entity called the "nominal compound" may be identified? I believe that in fact we must draw just such a conclusion. Although the literature on this creature is extensive and intriguing, all evidence suggests that the beast is mythical. It is true that we could arbitrarily define the "nominal compound" to be, let us say, just that configuration of two juxtaposed nouns which exhibits fronted stress; we would, however, have achieved little of linguistic significance thereby since we would have no way of predicting, much less explaining, the striking fact that all of the syntactic and semantic properties of our newly defined compounds are shared by many other forms which lack the morphological and phonological properties built into our "definition."

On the other hand, it surely cannot be the case that such distinguished authors as Jespersen, Bolinger, Marchand, and Lees were describing nothing more than a figment of their imagination. What I believe to be true instead is that these authors, and the many more who have studied "compounds," were, wittingly or not, selecting various subsets of complex nominals and restricting their analyses to just those forms and no others. Lees (1960), for example, is quite explicit in excluding from his study—"for convenience"—any forms with "secondary–primary stress" (i.e., "level" or "nonfronted" stress) although he freely states that "this distinction might be no more significant for the generation of nominal compounds than some orthographic difference [p. 119]." Other authors either failed to realize, or deliberately chose to ignore, the existence of the very extensive set of CNs containing nonpred adjs which parallel the N–N data in virtually every respect; although this very common restriction may make data selection and analysis seem easier, in most cases it has had the unfortunate result of obscuring larger generalizations and thus preventing the development of a more comprehensive and insightful theory.

It seems that such a state of affairs was made possible by the fact that the various sets of data examined do overlap to a considerable extent; this overlap provided investigators with apparently empirical support of their intuitive conviction that there **is** such an entity as the "nominal compound" and thus obscured the fact that no explicit definition had been provided anywhere that could distinguish "compounds" from "noncompounds" in a reliable, consistent, and systematic fashion. Unfortunately, this very strong intuitive conviction (shared, not incidentally, by this author for several years) may now be seen to reduce to nothing more rigorous than a fuzzy if comforting sense of the intersection of the various sets of citations encountered in the literature. To be sure, the examples most frequently cited **are** composed of two surface nouns, and are as likely as not to show fronted stress, permanent aspect, and semantic specialization. But surely frequency of citation in the literature, and familiarity through common usage, cannot stand as criteria for determining the grammatical description of a given form.

Has this investigation of the elusive "compound" been utterly in vain? Not entirely. We have learned and/or confirmed:

1. That the N–N forms in question cannot be isolated on any but the most superficial grounds from the comparable set of forms with nominal adjectives;
2. That some of these forms acquire fronted stress (under conditions which are not yet understood);
3. That semantic specialization accrues to many of these forms at a much higher rate than is true of "nominal phrases" in general.

These facts are not uninteresting, however little use they have proven to be in our search for the "nominal compound" in English; in fact, both 1 and 3 represent important aspects of the analysis of complex nominals presented in this book.

At this point, it is appropriate to return to the questions posed at the beginning of this section. Because we have failed to identify any valid and mutually consistent criteria for isolating "compounds" from the set of complex nominals, the problem of ascertaining the relationship of compounds to complex nominals becomes moot. Instead, we may now turn our attention to the latter construction and begin to sketch out the scope, direction, and content of a generative semantic analysis of the complex nominal in English.

3

TOWARD A THEORY OF COMPLEX NOMINALS

3.1 Introduction

This chapter is devoted to a number of fundamental issues that concern the area of complex nominals as a whole; the exploration of these issues will thus provide much of the theoretical foundation on which the detailed derivations of the following chapters are based. Section 3.2 begins with an outline of the major claims and premises of my theory, followed by a discussion of some of their more important theoretical implications. Particular attention is given to the controversy over whether CNs must be viewed as constructions which are so idiosyncratic as to preclude a formal analysis of their syntax and semantics, or whether they should be viewed instead as configurations which are multiply but predictably ambiguous and hence amenable to systematic linguistic study.

Section 3.3 then addresses the question, "Why do languages have complex nominals at all?" and offers answers based on communicative efficiency on the one hand, and the naming function of CNs on the other. In the last section, my claim that complex nominals must be analyzed syntactically as members of the category of nouns is supported by evidence drawn from a variety of syntactic phenomena; these arguments provide the theoretical underpinnings for one of the most important features of the derivations in Chapters 4 and 5, namely, the fact that CN formation operates in such a way as to create new nouns out of antecedent NP structures.

3.2 Major Claims and Premises

The fundamental claims and basic principles of my theory are listed here in outline form:

(3.1) 1. Complex nominals are all derived from an underlying NP structure containing a head noun and a full S in either a relative clause or NP complement construction; on the surface, however, the complex nominal is dominated by a node label of N.
2. Semantic restrictions are identical for complex nominals and for the propositions contained in their underlying structures, in the sense that they are no more and no less idiosyncratic for the former than for the latter.[1]
3. Any given CN form is inherently and regularly ambiguous over a predictable and relatively limited set of possible readings; although any one of these readings may be used more frequently than the others, or even exclusively, in a given speech community and during a certain period, the potential ambiguity still remains part of a speaker's competence and hence must be recognized in any grammatical description of these forms.
4. Complex nominals are all derived by just one of two syntactic processes: the **deletion** or the **nominalization** of the predicate in the underlying S.
5. For complex nominals derived by predicate **deletion,** a small set of Recoverably Deletable Predicates (RDPs) can be specified such that only its members, and no other predicates, may be deleted in the formation of CNs; the members of this set are CAUSE, HAVE, MAKE, BE, USE, FOR, IN, ABOUT, and FROM.
6. The potential ambiguity that is created by the multiplicity of possible underlying sources for a given surface CN is drastically reduced in discourse by both semantic and pragmatic considerations.

[1] In the present theory, this follows naturally and unsurprisingly from the fact that CNs are **derived** from fuller structures which include the propositions; nonetheless, this principle needs to be stated explicitly since many of the allegations of idiosyncrasy to be found in the literature appear to be a consequence of ignoring—or denying—the links between the broad range of semantic relations that can be expressed in sentential form and those that we find in compressed form in CNs.

. The relationships expressed by the RDPs appear to be of such semantic primitiveness that the set of RDPs proposed herein for English may well reflect universal constraints on the semantic structure of complex nominals in all languages; preliminary evidence from a number of other languages, as well as from other areas of the grammar, supports this hypothesis.

8. CN formation involves some processes which manifest the generality of regular, nongoverned syntactic transformations, and others which have more in common with the idiosyncrasies that characterize the areas of both morphological derivation and lexical entries; thus, an account of CN formation will differ in certain respects from accounts of fully general syntactic processes as well as from accounts of either partially productive derivational processes or totally idiosyncratic lexical entry rules.[2]

The reader will have observed from both the tone and the content of the principles outlined here that the present study is firmly founded upon the view that complex nominals are formed by grammatical processes whose regularities **and** complexities are amenable to systematic linguistic analysis in terms of both syntactic and semantic properties. This approach to the data has not, however, been uniformly adopted in previously published literature; on the contrary, linguists from Jespersen to Chomsky have considered various subsets of CNs (i.e., "nominal compounds," CNs with Proper Pseudo-Adjectives, NPs whose head nouns are nominalizations) to be so idiosyncratic as to preclude systematic formal analysis.[3] Let us consider here just why these linguists have maintained that no formal characterization is possible for these expressions, outside of the simple listing in the lexicon which is appropriate for idiosyncratic forms.

One of the major reasons offered in support of this allegation of idiosyncrasy is the fact that the semantic relationship of the prenominal modifier and the head noun is often not apparent from the surface elements alone; this may be seen by considering the variety of semantic distinctions that can be made for the CNs shown in (3.2)–(3.5).[4]

---

[2] This problem of the "intermediate" status of word formation processes in general (i.e., derivation as well as compounding) is addressed in more detail in Zimmer 1964. For a specifically lexicalist perspective on the subject, see Aronoff 1976.

[3] See, for example, Jespersen 1942, Volume 6, Chapter 8; Bolinger 1967:62–63; and Chomsky 1970, 1972a. A more detailed discussion of these three works in relation to the analysis of CNs may be found in Levi 1975, Section 3.1, where they are contrasted with a number of analyses based on ambiguity rather than idiosyncrasy, including those of Lees 1960, Gleitman and Gleitman 1970, and Coates 1971.

[1] The paraphrases supplied here have been constructed to highlight apparent idiosyn-

(3.2)   *musical clock*   = 'clock that makes music'
        musical comedy   = 'comedy that has music'
        musical interlude = 'interlude which is music'
        musical criticism = 'criticism of music'

(3.3)   *electrical clock*   = 'clock powered by electricity'
        electrical shock   = 'shock caused by electricity'
        electrical generator = 'generator producing electricity'
        electrical heating   = 'heating by means of electricity'

(3.4)   *tree branches*   = 'branches that a tree has'
        tree nursery   = 'nursery for trees'
        tree spraying   = 'spraying of trees'
        tree house   = 'house built in a tree'

(3.5)   *bedclothes*   = 'clothes for wearing in bed'
        bedpan   = 'pan for using in bed'
        bedroom   = 'room which contains a bed'
        bedpost   = 'post which is part of a bed'
        bedsore   = 'sore caused by lying too long in bed'

These data demonstrate that neither the head noun nor the prenominal modifier constitutes a completely reliable guide to the meaning of the CN as a whole. The nonpred adj *musical*, for example, appears to take on as many meanings as the head nouns it can modify, as may be seen from the variety of paraphrases in (3.2). On the other hand, keeping the head noun constant is no guarantee that the semantic relation between it and its modifier will remain constant either. For example, we normally understand *a musical clock* to be 'a clock that produces music' but *an electrical clock* is not 'a clock that produces electricity'. On the other hand, *an electric generator* does produce electricity but *a solar generator* certainly does not produce suns. With examples like these, the allegation of idiosyncrasy is at least understandable, even if not ultimately justified.

As we shall see in later chapters, it cannot be denied that there **are** idiosyncratic aspects to the grammar of complex nominals; these involve such factors as lexicalization, individual variation in the extent and organization of speakers' knowledge of CNs, historical remnants in contemporary English, and a certain analytic indeterminacy related to the fact that knowing an appropriate referent for a CN is not the same as

---

crasies comparable to those traditionally cited in the literature, and to represent the most common readings of these particular CNs (omitting some irrelevant details of semantic specialization); they thus do not reflect explicitly the multiple ambiguity that is in fact inherent in all CNs.

determining its grammatical derivation. But to conclude from these less tractable aspects of the subject that idiosyncrasy (especially, though not exclusively, in the semantic domain) is virtually a defining property of complex nominals is to make demonstrably untenable claims about the grammar of English and to ignore the weight of extensive evidence which in fact attests quite to the contrary.

Although the major part of this evidence belongs to chapters other than this one, two crucial arguments are well worth our attention here. These follow from the fact that an analysis of CNs in which each form would have to be learned individually could in no way account either for the universally acknowledged recursiveness of the CN formation process or for the spontaneous creation of new CN forms of which every native speaker is clearly capable. In a grammar in which CNs are treated as fundamentally idiosyncratic, and hence as elements to be listed with their definitions in the lexicon, the fact that any CN can be extended without theoretical limit (as in *apple pie, apple pie plate, apple pie plate crack, apple pie plate crack pattern, apple pie plate crack pattern studies*, and so forth) would entail the inclusion in the grammar of an infinite lexicon since every extended form would, as a matter of principle, have to be explicitly defined. The notion of an infinite lexicon is, however, internally inconsistent if we are to maintain our view that the lexicon is basically a finite **list** of just those unpredictable aspects of surface forms which can **not** be specified by some more general rule or rules.

Moreover, a theory which denies the existence of fundamental regularities in the formation of CN forms could not explain the fact that speakers freely and frequently create novel CNs (and extend the reference of already current CNs) without having to provide explicit definitions for these spontaneous linguistic creations. If there were indeed as little systematicity to the relationships expressible by CNs as some linguists have suggested, both listeners and readers in normal communicative situations would have no internal resources with which to interpret such unfamiliar forms or uses. This is, however, clearly not the case; in fact, it is precisely the ready interpretability of even novel CN forms (together with their expressive conciseness) that makes them such a widely used construction in English and other languages.

In summary, then, a grammatical description of CNs which aspires to completeness must certainly recognize the existence of certain idiosyncrasies in this area. There is, however, no justification for using these idiosyncrasies to obscure or deny the vast array of syntactic and semantic regularities that may in fact be discerned in the relevant data. It is, of course, just these regularities which this book is intended to describe.

It is important to recall at this point that my analysis of CNs ascribes not simply a pervasive systematicity to their grammar but, more specifically, a characteristic and specifiable ambiguity over a small set of potential readings for each form. As a consequence, one essential goal of my theory must be to predict any and all semantic structures that could regularly underlie a given surface expression, rather than any single possibility, however plausible that possibility may be. This point must be emphasized because familiarity with the most plausible or the most common reading of a CN often tempts us to describe that reading as "the meaning," when more careful analysis would show that the CN is in fact regularly and predictably ambiguous. For example, in ordinary discourse we would most likely use the CN *solar generator* to mean 'generator using [the energy of] the sun'; it is this reading which makes the most sense in view of our knowledge of what the "institutionalized referent" of this particular CN is (i.e., what object it normally names). However, it is equally true that the same CN could be used, given appropriate extralinguistic contexts, to mean 'generator which produces suns' (cf. *electric generator*), 'generator on the sun' (cf. *mountain generator*), 'generator for the sun' (cf. *emergency generator*), or 'generator having suns' (cf. *dry-cell generator*).

It is therefore essential that a complete theory of CNs be able to associate with a given surface form **all** the possible meanings which are permitted by the native speaker's competence, and not just the one that happens to be in common use. To be sure, in trying to understand others, we strive to disambiguate a given CN by whatever clues, linguistic and extralinguistic, that we can muster in order to select the one reading **intended** by the speaker. But a theory "whose goal is to correctly pair surface structures with semantic structures and contexts [McCawley 1973:236]" must not deny the inherent ambiguity but instead acknowledge and account for it. (This is not to say that the speaker's competence does not include the knowledge of which meaning is in common use, but this is a separate aspect of our competence—and a separate problem for any theory.)

It is thus my intention to provide a grammatical description of the open, synchronic productivity which characterizes CN formation; while this description will associate a number of semantic structures with a given surface CN, it will not and cannot predict which one of these will be used most frequently, much less whether the surface form is being used anywhere at all at a given point in time. (The failure of a grammatically possible CN to occur in the speech of a given individual or group of individuals could be the result of any number of unpredictable factors, including (*a*) simple ignorance on the part of some speakers of a form used in the speech of others; and (*b*) the equivalent of a "lexical gap" in the language that could be bridged at any moment by the introduction of a novel CN.)

It is important to distinguish here between two commonly confused senses of the word *productive*. As Zimmer (1964) points out:

> The term "productive" is often used rather indiscriminately to refer both to certain aspects of the behavior of the speakers of a language and to certain diachronic trends; while there is presumably in many cases a connection between these two aspects of productivity, it is necessary to keep the distinction in mind [p. 19].

When I characterize my analysis as a description of the "productive" aspects of CN formation, it is the former, synchronic sense of the word which is intended, a sense which expresses the open-endedness of our ability to form CNs rather than the relative frequency of a given derivational type of CN in our daily speech.

This qualification thus excludes the few remnants still found in our language of processes that have ceased to be "productive" in this synchronic, open-ended sense. For example, there is a small set of CNs (cited in Hatcher 1960) which seem to have been formed on the pattern of 'N_2 which reaches just to N_1'; this set includes *knee-pants*, *hip-boots*, *waistcoat*, and *breast-rail* (= 'the upper part of a balcony'). That this pattern is now unproductive may be seen by the impossibility in contemporary English of coining new forms such as **calf skirt*, **ceiling ivy*, or **shoulder hair* to express the same relationship. The fact that English contains CNs like *knee-pants* (as well as other synchronically unproductive compound patterns, such as the *pickpocket* type cited in Lees 1960:150–151) does not mean that a synchronic description must generate them; indeed, we must beware of letting such remnant forms skew our analysis simply because they are likely to appear in any reasonably extensive collection of familiar N–N forms. This methodological principle is hardly new; Brekle (1970:26) notes its enunciation many years ago by Brugmann, who stressed that it is the **process** of compounding itself as a creative linguistic act which must concern us, rather than any specific compound form or forms which historical accident may once have introduced into our language (Brugmann 1900:359ff.).

It is equally important to realize that the set of grammatical associations between meaning and form to be generated by my theory cannot be identified either with a set of "familiar" CNs or with any set of "attested" CNs, for any set of either familiar or attested CNs must necessarily comprise only a part of the set of grammatically possible CN expressions. For example, the creative interpretation of a CN like *vehicular engineer* as 'engineer with four wheels' (a reading supplied by a linguist–subject in a rudimentary psycholinguistic experiment of mine) cannot be rejected as "ungrammatical" on the grounds that the form itself is "unattested" or that the interpretation is an "unfamiliar" one; as Botha (1968) writes, "The unacceptability of compounds [whether judged by naive native speakers or by linguists] because of their unfamiliarity is no reason for not generating them [p. 149]."

To restrict the output of a theory of CN formation to only attested or familiar forms would be to deny the open-endedness of the processes involved, a denial which is demonstrably without empirical support. Moreover, any attempt even to specify which CNs are attested at a given point in the history of a language is doomed to failure since (*a*) there is no way to prove that the absence of a form is due to anything but the finite size of one's corpus of examples; and (*b*) the set of CNs in the repertoire of any individual speaker (or linguist) is limited to a high degree by extralinguistic factors such as education, social class, ethnic background, occupation, geographical location, and even hobbies, so that a CN which strikes one speaker as meaningless (and hence "unacceptable" or "impossible") may be fully comprehensible, transparent, and in frequent use for the next. For example, each of the following CNs is meaningful to me, although I would expect very few other English speakers to have **all** of these in their own repertoire: *milk silverware, bump set, zipper foot, sister node, tire iron,* and *barrel adjuster.*[5] On the other hand, a Buddhist chemical engineer who liked to sail and tend rosebushes on his days off would have many CNs in his repertoire that would be unfamiliar or uninterpretable to me. These distributional accidents are, however, not the kind of data that a grammatical description of CNs can be expected to account for; instead, we must formulate a theory to provide the broader framework within which the specific CNs in any individual's repertoire (itself constantly changing) can be appropriately produced and understood.

3.3 The Function of Complex Nominals

3.3.1 Communication in Compact Form

Before turning to the complex details of the derivation of complex nominals, we ought to ask the fundamental question: Why does the English language bother to produce this particular syntactic configuration? Why not leave well enough alone by preserving all underlying elements, and so avoid the multiple ambiguity, if not gross idiosyncrasy, so readily observed in these forms?

---

[5] These terms have referents in the spheres of Jewish dietary restrictions, volleyball, sewing machines, syntax, automobile tools, and bicycle parts, respectively.

A preliminary answer is suggested by the fact that prenominal modification appears to be a heavily exploited **syntactic target structure** for the English language;[6] that is, the "slot" which just precedes the head noun of an NP may be filled in English by modifiers which are derived from a tremendous variety of underlying sources. The examples of (3.6) illustrate the semantic and syntactic heterogeneity of this particular target structure; the reader will note that the nouns and nominal adjectives belonging to CN forms represent only two of many possibilities for prenominal modification.

(3.6)	N:	*divorce lawyer, apple pie*
	N + N (+ . . .):	*lab equipment factory (watchman . . .)*
	nom adj:	*polar regions, cardiac arrest*
	adv adj:	*former students, putative solutions*
	predicating adj:	*handsome princes, ugly frogs*
	active participle:	*dancing girls, laughing children*
	passive participle:	*stolen property, baked potatoes*
	adv:	*the then ruler, fast acceleration*
	adv + adj:	*slyly introduced diversions*
	adv + participle	
	1. active:	*scarcely breathing kittens*
	2. passive:	*thinly sliced salami*
	N + participle	
	1. active:	*fish-eating dinosaurs, fun-loving kids*
	2. passive:	*windblown hair, home-cooked meals*
	expletive:	*those goddam kids, that blasted bus*
	numeral:	*the seven dwarfs, the third estate*
	others:	*to-and-fro motions, anti-French sentiment, a call-it-what-you-like-but-I-call-it-chutzpah attitude*

The observation that prenominal modification appears to be a heavily exploited target structure in English is, however, more of a description than an explanation of this distribution; indeed, it only raises two more challenging questions, namely, **why** does the language conspire to produce this configuration as often as it does, and **how**—in the case of complex nominals—can it do so without critical loss of information?

The most obvious answer to the "why" question is compactness; by

[6] Binnick (1970) explains the notion of "syntactic target structure" as follows:
Although there are a great many types of underlying structures, there are relatively few surface structures . . . today we would talk of "syntactic conspiracies": over a whole derivation, transformations literally cooperate to transform a whole variety of structures into just certain surface configurations [p. 241].

the formation of complex nominals, a head noun and its underlying proposition can be reduced to just two words. As Vendler (1967) remarked of nominalizations (which form the head nouns of a large subset of CNs), "They are a means of packing a sentence into a bundle that fits into other sentences [p. 125]."[7] CN formation is therefore a device of tremendous utility in English, if only for expanding the amount of information we can store in short-term memory.[8] But this observation leads directly to the "how" question previously posed: How can this compression be accomplished without losing essential semantic or syntactic information?[9]

To begin with, the **semantic** relationship of modification is preserved by the **syntactic** configuration of prenominal element plus head noun; a glance at the wide variety of prenominal modifiers possible in English, illustrated in (3.6), indicates that all prenominal elements, regardless of their morphosyntactic composition, are interpreted as modifiers of the last noun in the NP. And since this head noun is always the *déterminé* and the prenominal element the *determinant*, the endocentricity characteristic of all regularly formed CNs is systematically preserved by the simple syntactic device of word order.[10] (Of course, in languages where the head noun normally precedes its modifiers, we would expect to find CNs in which the *déterminé* precedes the *déterminant*; this is borne out by French CNs such as *une phrase noyau*, *une guerre-éclair*, *la décision présidentielle*, and Hebrew CNs such as *aruxat boker* 'meal' + 'morning' = 'morning meal, breakfast' or *yeled kibuc* 'child' + 'kibbutz' = 'kibbutz child, child of the kibbutz'.)

---

[7] In point of fact, neither nominalizations nor CNs of any type are formed from "bundles" containing only sentences, as Vendler here suggests. Although this misconception is found fairly frequently in the literature (e.g., in Brekle 1970, Downing 1975, Marchand 1969, and Reif 1968), it is neither semantically nor syntactically defensible. As will be shown in detail in Chapters 4 and 5, both nominalizations and CNs in general must be derived from source structures which comprise both a head noun **and** a sentential complement. Vendler's more general point about the conciseness achieved by nominalizations still obtains, however.

[8] Downing (1975) correctly points out that it is just this feature of conciseness (together with the naming function of CNs) that explains the heavy use of compound forms in newspaper headlines: "Compounds are ideally suited for use in situations where there is a premium on brevity, yet no appropriate unitary item exists. Thus, newspaper headlines, etc., are filled with compounds . . . [p. 45]." Jespersen (1942:137) and Brekle (1970:190) have also commented on this abbreviatory function of "compounds."

[9] For a moralistic tale which focuses on just this question, see Levi 1974:412–413.

[10] As noted earlier in Chapter 1, exocentric CNs may be regarded as regularly formed CNs whose head nouns have been lost by the process of "beheading," a process which destroys the otherwise typical endocentricity of the original form. For example, we may surmise that an exocentric form like *birdbrain* is derived from an antecedent structure something like **birdbrain haver*, which is as endocentric and regular a CN as, say, *home owner* (except for the fact that **haver* never surfaces in that particular form).

The second reason that CN formation can take place without critical information loss is that either (*a*) all specified elements in the underlying structure are realized overtly (as in nominalizations like *senatorial industrial investigation* from, roughly, [ACT *## senators investigate industry ##*]); or (*b*) an underlying constituent which has been deleted is nonetheless recoverable in a systematic fashion (as in cases like *drug deaths* from *deaths* CAUSED BY *drugs*). The mechanism by which such material is recovered from the compressed CN configuration will be explained in detail in the chapter to follow. For the present, it suffices to observe that as a result of the constraints on CN formation to be proposed, the removal of a meaning-bearing element in the course of the derivation of a CN produces not an irrecoverable loss of semantic information, but rather that multiple and predictable ambiguity over a specifiable range of readings that has already been outlined in earlier pages. It is because of these constraints that the processes which form CNs may operate, productively and recursively, without losing crucial amounts of semantic information.

These comments pertain particularly to CNs derived by predicate deletion. In the case of CNs derived by nominalization, a different mechanism contributes to the characteristic concision of these forms, namely, the deletion of unspecified object NPs. Thus, we regularly find CNs like *presidential denial*, *parental request*, or *city regulations*, where the surface form fails to tell us just what was denied, requested, or regulated. What makes this deletion particularly significant is that it is permitted in CN formation but prohibited in the full sentences that are part of the source structures from which the CNs are derived. This asymmetry is displayed in (3.7) and (3.8); in both sets of examples, the nonnominalized verbs must have an overtly expressed object, while the same verbs, when nominalized within CNs like those in (3.8), may appear without any object NP.

(3.7) a. *The virus infected* $\begin{Bmatrix} the\ cattle \\ {}^{*}\varnothing \end{Bmatrix}$.

b. *The revisionists betrayed* $\begin{Bmatrix} their\ comrades \\ {}^{*}\varnothing \end{Bmatrix}$.

c. *The nation consumed* $\begin{Bmatrix} many\ nonrenewable\ resources \\ {}^{*}\varnothing \end{Bmatrix}$.

(3.8) a. *He was afraid* $\begin{Bmatrix} of\ a\ viral\ infection \\ {}^{*}that\ a\ virus\ would\ infect \end{Bmatrix}$.

b. *She predicted* $\begin{Bmatrix} a\ revisionist\ betrayal \\ {}^{*}that\ the\ revisionists\ would\ betray \end{Bmatrix}$.

c. *Economists were surprised by the extent*
$$\left\{\begin{array}{l}\textit{of national consumption}\\ \textit{*to which the nation consumed}\end{array}\right\}.$$

These data thus illustrate a third way in which CN formation preserves all essential information despite the compactness of form; no semantic content is lost in the deletions of (3.8) simply because the object NPs were never specified in semantic structure. (Of course, objects that are specified in LS must surface in some form or other, whether in an "objective genitive" phrase such as *revisionist betrayal of the proletariat*, as a preposed noun, as in *national oil consumption*, or as a nominal adjective, as in *industrial electrical production*; the asymmetry in deletion is manifested only with respect to unspecified NP objects.)

This deletion phenomenon illustrates the only area of syntax cited by Ross (1972a) as reversing the "category squish" or "funnel direction" which in all other respects demonstrates that "nouns are more inert, syntactically, than adjectives and adjectives than verbs [p. 325]." Ross continues:

> The one case I know of in which the "funnel direction" appears to be reversed concerns the degree to which constituents which appear in remote structure must appear at certain later levels of structure (say, at the output of a cycle). The relevant facts appear in (28):
>
> (28) Constituent deletion: the nounier, the fewer constituents need appear in surface structure [p. 325].

What remains unexplained here is why the "category squish" noted by Ross should reverse itself in this particular area; specifically, why should the process of nominalization render unspecified direct objects so readily deletable in CN form? The valid observation that such deletability helps to produce the communicative brevity of CNs only raises the question as to why such brevity may not also be exploited within full sentences. Unfortunately, this observation (like Ross's observation concerning the existence of the more general "category squish") constitutes a description of the facts but not an explanation; for the latter, we must await further research.

3.3.2 The Naming Function

We have just seen, albeit briefly, that the operation of certain syntactic and semantic constraints permits the process of CN formation to achieve a distinctive compactness of form without drastic loss of semantic content. We may, however, take our inquiry one step further by trying to ascertain in a more general way the function of these forms in human communication. This has, in fact, been the purpose of both Zimmer (1971, 1972) and Downing (1975, 1977), whose functional approach to nominal compounds will be briefly reviewed here.[11]

Zimmer's 1971 work is the earliest attempt (of which I am aware) to shift the attention of linguists away from their usual focus on attested and familiar compounds to the area of new compounds and the conditions under which they are coined. It is Zimmer's claim that the major function of nominal compounds is as **naming devices** which pick out particularly salient categories of the speaker's experience. For example, although the transitory shape of a cloud may be of little interest to most speakers of English, it could have "classificatory relevance" to members of a sect which interprets animal-shaped clouds as omens of great significance; Zimmer thus claims (1971:C14) that the sight of a cloud shaped like a kangaroo might lead speakers of the former group to utter (3.9a), but that only speakers of the latter group would be led to coin a new compound by uttering (3.9b).

(3.9)   a. *Hey, there's a cloud that looks like a kangaroo!*
          b. *Hey, look at that kangaroo cloud.*

The hypothesis that "the naming function of compounds . . . is based on the potentially classificatory nature of the relation between their constituents [Zimmer 1971:C16]" helps to explain why different speakers might coin different compounds to describe the same referent, or, as in the *kangaroo cloud* example, why some might use no compound at all but simple description instead. In discussing these variables, Zimmer (1971) writes:

> The dimension of classificatory relevance that I am trying to define here has something to do with the distinction between naming and description. Anything at all can be described, but only relevant categories are given names (I am talking here about common rather than proper names) [C15].

Expanding on the work of Zimmer, Downing (1975) conducted several experiments in which subjects were asked (*a*) to coin one- or two-word "descriptions" for unfamiliar objects depicted in drawings, (*b*) to supply interpretations for novel CNs presented out of context, and (*c*) to judge

---

[11] Since Zimmer and Downing have both written in terms of "nominal compounds" rather than "CNs," I will use the two terms interchangeably in this section to avoid clumsy restatements of the obvious; this stylistic concession, however, in no way invalidates the conclusions of Section 2.5 with respect to the status of "nominal compounds."

the degree of plausibility of a variety of interpretations provided for particular novel CNs. As a result of these experiments (and of her analysis of previous work in the area of compounding), Downing concludes that a speaker who is led to create a novel compound

> is typically faced with a situation in which he wishes to refer to an entity which possesses no name of sufficient specificity for his classificatory or communicative purposes. Because the compounding process is extremely productive, and because compounds are considerably more transparent semantically than novel monomorphemes, *compounds are ideally suited to serve as ad hoc names* [Downing 1975:11; emphasis added].

Thus, both Zimmer and Downing emphasize the difference in function between phrases which **describe,** and compounds which **name;** in addition, Downing notes that the naming function of compounds should also be distinguished from the **asserting** function performed (she claims) by sentential paraphrases of compounds (Downing 1975:42).[12]

These distinctions are important in that they suggest a way of explaining why only some underlying structures are changed into CN form, while others reach the surface without being so compressed; if Zimmer and Downing are correct, only those antecedent structures which are needed for names, and which express a relation of "classificatory relevance" for the speaker (and presumably for the addressee as well), will be transformed into novel CNs. (Although the use of already institutionalized or lexicalized CNs is not at issue here, we may note that the lexicalization of a given CN itself constitutes evidence for its "classificatory relevance" within the speech community in which it is used.)

Whereas both Zimmer and Downing have focused on the creation of novel compounds, and the conditions that trigger such an event, it is obvious that many CNs not only survive the moment of their birth but flourish long after; indeed, most of the examples cited in this work represent such CNs, whose more or less long life may be taken as evidence of their utility in naming particular referents. The fact that so many CNs achieve this longevity (often accompanied by semantic specialization) is aptly reflected in Downing's remark that "compounding

[12] While it is certainly true that CNs name rather than assert, it is not true that CNs are adequately paraphrased by sentences alone, as Downing claims here. As noted in footnote 7, we may appropriately derive a CN only from a head noun **and** a sentential complement, within an overall structure dominated by an NP. With such a source structure, it then follows naturally that CNs would share the naming function characteristic of nouns or noun phrases, rather than the asserting function characteristic of (some) declarative sentences.

. . . serves as a back door into the lexicon [Downing 1975:45]." This is particularly true when we consider not only the high frequency with which certain CNs are used, but also the fact that a CN which persists well beyond its first coinage "acquires more and more of the characteristics of a unitary lexical item [Downing 1975:45]." It seems that Downing is here referring to the fact that CNs, once coined, are subject to the process of lexicalization (whether this involves simply the "institutionalization" of one reading out of the range of possible readings, or semantic specialization above and beyond any such reading), as well as to the forward shift in stress that distinguishes many "compounds" from "phrasal expressions." In addition, we can reasonably hypothesize that many high-frequency CNs are neither initially learned nor ever perceived as the complex transparent output of a regular syntactic process but are instead entered in a speaker's lexicon as single, unanalyzed units. (Such a hypothesis raises certain questions about the validity and productivity of a grammatical model of CN formation such as the one proposed in these pages. For discussion of this issue, see Section 7.3.)

Downing's characterization of CN formation as "a back door into the lexicon" is particularly appropriate when we consider the changes that specific CNs often undergo over time. That is, while CNs may in many cases be appropriately analyzed as the output of a regular, transformational process, like the CNs in (3.10), in other cases CNs that might once have been so analyzed have become semantically specialized to the extent that their specific referents are no longer predictable from the surface constituents, as may be seen by the CNs in (3.11).

(3.10)   *mountain cabins, mountain temperatures, mountain clothing, mountain vacation, mountain villages, mountain warfare*

(3.11)   *mountain lion* [= *puma*], *mountain laurel* [= *a specific plant*], *mountain beauty* [− *a kind of trout*], *mountain cow* [= *tapir*], *mountain witch* [= *a kind of dove*], *mountain flour* [= *a mineral*]

A lexicographer producing a dictionary of English would be ill advised to attempt to list transparent CNs like those in (3.10) (if only because the list is in principle limitless—see Section 6.2), but would be correct in providing entries for CNs like those in (3.11) which have acquired special meanings. (In fact, all of the latter are entries in the Oxford English Dictionary [OED], along with many more animal, plant, and mineral names formed with *mountain* as a prenominal modifier.)

It is instructive to view this difference in the degree of semantic specialization in terms of location along a **continuum of derivational**

transparency, such that the CNs in (3.10) would be located closer to the "transparent" pole of the continuum, whereas the CNs in (3.11) would be situated closer to (though not at) the "opaque" pole. If we limit our focus to CNs alone, such a continuum might have entries arranged in the vertical sequence indicated in (3.12).

(3.12)   Transparency
           a. *mountain village, family reunion, lemon peel,* . . .
           b. *grammar school, briefcase, polar bear, boy friend,
              ball park,* . . .
           c. *birdbrain, razorback, cottontail, blockhead,* . . .
           d. *polka dot, monkey wrench, flea market,* . . .
           e. *honeymoon, fiddlesticks, duck soup,* . . .
           Opacity

The highest entries here are those which need not be listed in a dictionary (and, correspondingly, which I claim are derivable by regular syntactic processes); the next entries are those which probably once were transparent (such as *grammar school* or *briefcase*) but have since become more opaque, by either semantic narrowing or broadening. The third group (c) contains exocentrics, whose surface components are combined according to the rules of regular CN formation. Here, however, the intended meaning is obscured on the surface by the fact that the head noun has been lost by "beheading"; for these superficially regular forms, then, the identity of the head noun must be learned as an idiosyncratic feature (e.g., whether a *razorback* is animal, mineral, or vegetable). The fourth and fifth groups contain CNs which are partially or wholly idiomatic, respectively, and hence represent the two most opaque kinds of complex nominals.

The notion of a continuum of derivational transparency is also instructive in regard to a number of other aspects of CN formation. To begin with, we may note that the CNs ranged closer to the opaque pole in a synchronic description may have moved there gradually (as I surmise may be true for *briefcase* or *grammar school*, whose meanings appear to have broadened from their original senses) or may have been there from their initial coinage (as must have been true for *mountain beauty*, used to denote a certain kind of trout, or *ladyfinger*, a certain kind of pastry). To extend Downing's metaphor, some CNs may have crept slowly through the back door of the lexicon whereas others may have crossed the threshold all in one leap.

In discussing historical processes, it is also interesting to observe that many lexical items derived from Greek or Latin roots may be regarded as "etymological" compounds, whose noun–noun structure must have been obvious to the speakers who first coined them and whose composition will be transparent even today to speakers with sufficient education or interest to have learned their classical origins. Consider the parallels in (3.13).

(3.13)

	a. "Etymological" CNs	b. "Synchronic" CNs
	hippopotamus	*sea horse*
	theology	*divinity studies*
	cuneiform	*heart shape*
	thermometer	*water meter*
	aqueduct	*water conduit*

Along the derivational continuum we are considering, the CNs in (3.13a) would of course be ranged closer to the opaque end than the CNs in (3.13b); nonetheless, the former would be further from the pole of opacity than monomorphemic forms which show neither synchronic nor historical derivational "seams."

To summarize these observations, the forms produced by CN formation can remain outside the lexicon of a given speech community just so long as they retain their derivational transparency; however, once they acquire idiosyncratic elements of meaning, they must be considered to have entered the lexicon that the next generation will have to learn, and thus will have moved down the derivational continuum in the direction of increased opacity.

As a result of many arbitrary factors (for there are certainly many "back doors" into the lexicon in addition to compounding), the degree of derivational transparency within a set of semantically related items will vary widely; in addition, individual speakers will have varying degrees of awareness concerning the derivational composition of both "etymological" and "synchronic" CNs. A final illustration of these points is given in (3.14), where leftmost position marks the most opaque lexical items.

(3.14)

Surface monomorph	Etymological polymorph	Lexicalized CN	Nonlexicalized CN
kitchen	*parlor*	*bathroom*	*game room*
horse	*hippopotamus*	*sea horse*	*farm horse*
stable	*aviary*	*lighthouse*	*staff house*

In this section, we have examined the function of complex nominals from a number of different perspectives. On a relatively simple structural level, we have seen that CNs illustrate one type of prenominal modification, whose formation appears to reflect the strength of "attraction" of that particular target structure of English. From another point of view, CNs may be considered as a very successful means of compressing semantic and syntactic information into a highly compact form. Finally, seen from the functional perspective developed by Zimmer and Downing, complex nominals can be understood as naming devices that pick out the "relevant categories" of a speaker's experience. Two factors which make CNs so efficient as naming devices are (*a*) the semantic transparency with which most CNs begin their life, and (*b*) the tremendous productivity with which they may be generated. On the other hand, we must also recognize that many CNs sooner or later lose their initial transparency as they acquire elements of specialized meaning; these forms then enter the lexicon and so must be learned as idiosyncratic units by the next generation.

3.4 Category Membership of Complex Nominals[13]

In their absorption with the problems of delineating the internal relations (whether semantic or syntactic) of nominal compounds, linguists have paid remarkably little attention to the external relations of these forms with other parts of the sentence. More specifically, while most linguists appear to have **assumed** that nominal compounds should themselves be analyzed syntactically as nouns (cf. Lees 1966, Chomsky and Halle 1968, Jackendoff 1975), very little syntactic evidence has been offered in support of this assumption, nor have the ramifications of this claim been pursued in a systematic fashion.

Moreover, it has not been appreciated that the very same reasons for suggesting that nominal compounds are nouns in surface structure may be advanced to argue that other types of CNs are also surface nouns, namely, CNs with nonpredicating adjectives like *avian marauders* or *traumatic arthritis,* and CNs derived by nominalization such as *parent(al) interference* or *moon/lunar exploration.* The failure to realize the relatedness of these forms, with respect to category membership as well as many other features, has in turn led to the failure to recognize the degree to which these forms interact with each other to produce additional CNs. But if we claim that all these forms are to be treated as nouns themselves, it follows from the universally acknowledged recursiveness of the "compounding" process that any of these three types may interact with any of

---

[13] Much of this section appeared first in Levi 1977.

the others—and that the result must itself be a noun. Thus, the claim that
apple pie is a noun can be shown to lead, step by step, to the claim that
the forms in (3.15) must also be nouns; this latter claim, however, has not
been made anywhere in the literature despite its being one of the more
important ramifications of the initial assumption.[14]

(3.15)   a. *national vehicular control standards enforcement problems*
          b. *lunar exploration project soil molecular analysis equipment
             failure*
          c. *senatorial national export advisory board investigation cover-
             up scandal*
          d. *electrical engineering night school instructional materials
             editorial board salary increase conference agenda*

The purpose of this section is to provide the syntactic argumentation
that is required to demonstrate that all of these forms (i.e., CNs of all
types) must be regarded as tokens of the same surface syntactic category,
namely, the category of noun. The arguments will be based on evidence
from affixation, ordering relationships with determiners and adjectives
within the same NP, and anaphoric island phenomena.

We may note, to begin with, that traditional arguments for assigning
or determining category membership are most often based on the inflectional patterns of the language. Since, however, the relevant inflectional
endings in English (plural and possessive /s/) would be the same for both
Ns and NPs (the only two relevant syntactic categories in current generative semantic theory), we must turn instead to derivational affixes to
provide us with pertinent arguments for distinguishing between these two
types of nominal constituents. The claim that CNs are Ns rather than
NPs predicts that derivational affixes attachable solely to the former but
not to the latter should also be attachable to CNs; the data in (3.16)–(3.19)
show that this is indeed the case.

The data in (3.16) show the distribution of the English prefixes *post-*,
ex-, *anti-*, and *non-*; the three forms in each set of data indicate that the
prefix in question may be adjoined to both simple nouns and CNs, but
not to full NPs.[15] The evidence is most clear when the NP being tested

[14] In Levi 1975 I argued that all the cited forms must be viewed as syntactically similar,
thus recognizing explicitly that the forms in (3.15) must share the same constituent label as
simpler forms like *apple pie*; I erred, however, in my earlier research in describing these
forms as NPs rather than Ns. In correcting that error, this section should also make clear
why the expression **nonpredicate NP** which appeared in Levi 1973, 1974, 1975 has been
replaced here by **complex nominal.**

[15] These four prefixes are not intended as an exhaustive selection; other prefixes for
which parallel data could be cited include *pro-*, *pseudo-*, and *mini-*, as in *pro-
transformational grammar*, *pseudo-arms control*, and *mini-energy crisis*.

includes an article, as in *the post-[a sudden death] confusion* (= 'the confusion after a sudden death'), but the generalization holds as well even when articles play no part, as in *ex-[heavily drinking executives]*. Brackets are included here and throughout this section to clarify the constituent structure visually; there can be no doubt, however, that an *ex-copy editor* is not an 'editor of ex-copy', or that the *post-Arab summit period* does not refer to a 'summit period which is post-Arab'.

(3.16)   a. *the post-[war] years*
         the post-[Arab summit] period
         **the post-[a sudden death] confusion*

         b. *an ex-[husband]*
         an ex-[copy editor]
         **ex-[heavily drinking executives]*

         c. *anti-[war] demonstration*
         anti-[gun control] demonstration
         **anti-[the government's intervention] demonstration*

         d. *non-[Chicago] accent*
         non-[Hebrew speaker] accent
         **non-[(a) competent speaker] accent*

Evidence from English suffixation may be used in a similar fashion to demonstrate that CNs have the syntactic properties of Ns rather than NPs; this will be done here by considering the agentive suffixes *-ist* and *-ian*, as well as the adjectivalizing suffix *-(ic)al*.[16] The agentive data are illustrated in (3.17).

(3.17)   *[political cartoon]ist, [biological scient]ist, [social satir]ist, [molecular physic]ist, [polyphonic music]ian, [criminal law]yer, [generative grammar]ian, [historical lingu]ist, [critical essay]ist, [quantum mechanic]ian, [symbolic logic]ian*

Note that alternative constituent analyses are possible for some of these expressions, such as *social satirist*, which could arguably be analyzed either as *[social] [satir-ist]* 'satirizer of society' (an objective nominalization) or as *[social satir]-ist* 'one who does [social satire]'; as a result, these forms do not constitute conclusive evidence for the argument at hand (although they are compatible with it). For some of the forms in (3.17), however, the first of these alternative analyses is excluded on semantic

---

[16] That other suffixes work similarly, if less productively, is suggested by the colloquial expressions *[women's libb]er* and *[civil rights]nik*, as well as by such technical terms as *[anaphoric island]hood* or *[sentential subject]ness*.

grounds, and it is these which force us to accept the claim that it is the CN as a whole to which the suffix is added, rather than just the head noun.

A representative sample of the crucial forms is given in (3.18), where we see from the bracketing that here, too, the agentive suffixes in question may be attached to stems which are either Ns or CNs but not to NPs in general (i.e., not to NPs containing more than just an N or CN). (In each case, the reading intended for each expression is of the form 'one who does x' where the suffix expresses 'one who does' and the x is the nominal expression shown in brackets.)

(3.18)

a. *Al's a* $\begin{cases} [grammar]ian \\ [Montague\ grammar]ian \\ {}^*[manifestly\ internally\ inconsistent\ grammar]ian \end{cases}$.

b. *Lee's a(n)* $\begin{cases} [histor]ian \\ [intellectual\ histor]ian \\ {}^*[overquantified\ histor]ian \end{cases}$.

c. *Jo's a* $\begin{cases} [lingu]ist \\ [historical\ lingu]ist \\ {}^*[needlessly\ complex\ and\ overformalized\ lingu]ist \end{cases}$.

d. *Jeff's a* $\begin{cases} {}^*[mechanic]ian \\ [quantum\ mechanic]ian \\ {}^*[long\text{-}outdated\ mechanic]ian \end{cases}$.

The bracketing indicated in (3.18) is required by semantic considerations. For example, a *Montague grammarian* does not mean 'a grammarian dealing with Montague' but only 'one who does Montague grammar'; similarly, a *historical linguist* is not 'a linguist dealing with history' but rather 'one who does historical linguistics [= history of language(s)]'.

The clearest case of all is that of (3.18d). It happens to be the case that physicists who specialize in statistical mechanics, quantum mechanics, or (in earlier days) classical mechanics may properly be called *statistical mechanicians, quantum mechanicians,* and *classical mechanicians,* respectively; physicists, however, never use the word *mechanician* in isolation.[17] We must conclude, therefore, that the phrase *quantum mechanician* in (3.18d) is derived not by combining the nonexistent noun *mechanician* with the noun *quantum,* but rather by combining the full CN *quantum mechanics* with the appropriate agentive suffix. Note, however, that the possibility of effecting such a combination is exactly what is predicted by the hypothesis that CNs are in fact nouns.[18]

[17] My thanks to an astrophysicist friend, Jeffry Mallow, for this bit of evidence.

[18] We may note in passing that a similar argument could be adduced for adverbial expressions like *quantum mechanically, truth functionally,* and *set theoretically.*

The highly productive adjectivalizing suffix *-(ic)al*, normally added to
Ns (as in *musical, national, industrial*), is also found attached to a
relatively small number of CNs within the domain of technical discourse;
typical examples are expressions like *set theoretical framework, ego psychological analysis*, or *text grammatical models*. The reasons for the
much more restricted use of this suffix on CNs than on "simple" Ns, even
within the realm of specialized language, will be discussed later on in
Section 4.2.4, in the context of the Morphological Adjectivalization rule
which produces these forms. For present purposes, however, it suffices to
observe that the distribution of the suffix *-(ic)al*, as illustrated in (3.19),
shows that it too appears on both Ns and CNs but not on NPs of any
greater complexity.

(3.19)

a. *I dislike that* $\left\{\begin{array}{l} [theoret]ical \\ {[set\ theoret]ical} \\ {}^*[antiquated\ theoret]ical \end{array}\right\}$ *framework.*

b. *He's overfond of* $\left\{\begin{array}{l} [psycholog]ical \\ {[ego\ psycholog]ical} \\ {}^*[distorted\ Freudian\ psycholog]ical \end{array}\right\}$ *analyses.*

c. *These examples all show* $\left\{\begin{array}{l} [electric]al \\ {[left\ peripher]al} \\ {}^*[VPs\ in\ German]ic(al) \end{array}\right\}$ *gapping.*

d. *Have you tried a(n)* $\left\{\begin{array}{l} [epistemolog]ical \\ {[quantum\ mechanic]al} \\ {}^*[African\ atonal\ music]al \end{array}\right\}$ *interpretation?*

To summarize the affixation evidence, we have seen that the data
presented in (3.16)–(3.19) support the claim that CNs should be treated
syntactically as nouns; both the prefixes in (3.16) and the agentive suffixes
in (3.17) may be regularly attached to Ns and CNs, but not to NPs. The
adjectivalizing suffix in (3.19) also follows this pattern; moreover, the fact
that its use on CNs is highly limited does not reduce its significance as
evidence, since that limitation can be explained by an independently
motivated pragmatic constraint to be discussed in Section 4.2.4. A theory
which analyzed CNs as NPs, or even as some nonconstituent subpart of
an NP, could not predict the data on affixation without including some ad
hoc principle for distinguishing those "NPs" that could take N affixes
(i.e., CNs) and those that could not; in contrast, all the affixation data just
presented follow naturally and without descriptive duplication in a theory
that includes CNs within the surface category N.

Important evidence for the nounhood of CNs may also be drawn from the distribution of adjectives and determiners within different kinds of NPs. Taking adjectives (of the predicating kind) first, we observe that they may not be interposed between the two components of a CN; this fact follows, however, from a theory in which CNs are Ns, since predicating adjectives always precede nouns within an NP and never interrupt them. Some pertinent data are provided in (3.20) for a variety of CN types.

(3.20)   a. *the* $\begin{Bmatrix} \textit{pope's urgent} \\ \textit{urgent papal} \\ \textit{*papal urgent} \end{Bmatrix}$ *appeal*

   b. *extreme* $\begin{Bmatrix} \textit{thermal} \\ \textit{heat} \end{Bmatrix}$ *stress*

   * $\begin{Bmatrix} \textit{thermal} \\ \textit{heat} \end{Bmatrix}$ *extreme stress*

   c. *isolated* $\begin{Bmatrix} \textit{avian} \\ \textit{bird} \end{Bmatrix}$ *sanctuaries*

   * $\begin{Bmatrix} \textit{avian} \\ \textit{bird} \end{Bmatrix}$ *isolated sanctuaries*

   d. *the* $\begin{Bmatrix} \textit{government's recent} \\ \textit{recent government(al)} \\ \textit{*government(al) recent} \end{Bmatrix}$ *initiatives*

In these examples, we observe that the predicating adjective must in every instance precede the entire CN, even when the first element of that CN is itself an adjective; thus the CN *thermal stress* is as inviolable a unit as the synonymous form *heat stress*. One additional fact of interest to note here is that paraphrases of CNs which contain genitival modifiers rather than nominal adjectives or nouns (e.g., *the pope's* instead of *papal*, or *the government's* instead of *government*[*al*]) may not be analyzed as CNs, since the genitival forms do in fact permit the interposition of other modifiers between themselves and the head noun, as in *the pope's urgent appeal* and *the government's recent initiatives* in (3.20). (We will return to this distinction in Chapter 5, where the rules that produce these two syntactic and stylistic variants are examined in more detail.)

The distribution of determiners within NPs provides additional support for treating CNs as nouns. In English, nouns which are grammatically singular, count, and common do not constitute full NPs by themselves but must instead be preceded by a determiner of some sort (whether this be an article, a quantifier, or a genitival phrase). This requirement is demonstrated by the data in (3.21), where we also see that exactly the same distribution obtains for grammatically comparable CN forms.

(3.21)

a. *The/A/Any/*Ø* $\left\{\begin{array}{l} contribution \\ faculty\ contribution \\ parental\ contribution \end{array}\right\}$ *will be welcome.*

b. *The/A/Carnegie's/*Ø* $\left\{\begin{array}{l} sanctuary \\ bird\ sanctuary \\ avian\ sanctuary \end{array}\right\}$ *was established in 1925.*

c. *The/Any/Newark's/*Ø* $\left\{\begin{array}{l} planner \\ city\ planner \\ urban\ planner \end{array}\right\}$ *faces many obstacles.*

If CNs were dominated by NP nodes, they could function as subjects by themselves; the data in (3.21), however, indicate that CNs do not so function but instead require determiners in exactly the same way that noncomplex nouns do.

A particularly intriguing source of support for treating CNs as nouns is the set of data showing that CNs constitute **anaphoric islands** in the sense of Postal 1969. In that article, Postal claims that compounds, like other forms he terms "derivatives" (see Postal 1969:213), behave just like monomorphemic lexical items with respect to both inbound and outbound anaphora of various types. Specifically, he claims that no single component of a compound can act as antecedent for an anaphor elsewhere in the sentence, nor can a compound form include within it an anaphoric device (i.e., an "inbound anaphor").

The data in (3.22)–(3.25) appear to verify Postal's claim. In each case, the (a) sentence contains a well-formed anaphor whose antecedent is an NP consisting of a single noun; however, once that noun becomes part of a CN (as shown in the synonymous [b] sentences), the grammaticality of the resultant sentence is destroyed. (In the examples here, a referential index on a nominal adjective is to be interpreted as the index assigned to its underlying noun.)

(3.22)   a. *Lawyers dealing in divorce$_i$ often experience it$_i$ themselves.*
      b. **Divorce$_i$ lawyers often experience it$_i$ themselves.*

(3.23)   a. *Engineers specializing in electricity$_i$ have to study it$_i$ for years.*
      b. **Electrical$_i$ engineers have to study it$_i$ for years.*

(3.24)   a. *Drivers of trucks$_i$ spend days at a time in them$_i$.*
      b. **Truck$_i$ drivers spend days at a time in them$_i$.*

(3.25)   a. *Criticizing by parents$_i$ may alienate their$_i$ children.*
      b. **Parental$_i$ criticizing may alienate their$_i$ children.*

It would be difficult indeed to explain this change in anaphoric accessibility except by means of a distinction in constituent structure; but a theory which characterizes CNs as Ns provides detailed support for just such a distinction. The derivations in Chapters 4 and 5 will make explicit just how the "autonomous" NPs that are input to the CN formation process lose their autonomy (and thus their anaphoric accessibility) when they are adjoined to form a CN; for present purposes, it will suffice to note that the change is basically one which transforms a syntactic configuration of independent constituents into the much more cohesive, and hence more impenetrable, structure of a single noun. This fundamental insight of Postal into the anaphoric parallelism between monomorphemic lexical items and the set of "derivatives" which includes complex nominals thus provides an additional argument in favor of treating CNs as Ns.[19]

We have now seen that evidence from affixation, adjective and determiner distribution, and—to a more limited extent—anaphoric island phenomena provides systematic justification for the claim that CNs must be analyzed as syntactic nouns. To this may be added one more argument: that of descriptive simplicity.

In order to write the rules for forming the simplest kind of CN (i.e., a two-noun "compound"), one must require that the element which will emerge on the surface as the prenominal modifier be a single, unmodified noun, rather than just any NP. This requirement permits us to generate (3.26a), for example, but not the ungrammatical forms of (3.26b–e).

(3.26)   a. *bird sanctuary*
          (< *sanctuary* FOR *birds*)
      b. **lost-or-tired-bird sanctuary*
          (< *sanctuary* FOR *lost or tired birds*)
      c. **any-birds-at-all sanctuary*
          (< *sanctuary* FOR *any birds at all*)
      d. **all-those-damned-birds sanctuary*
          (< *sanctuary* FOR *all those damned birds*)

[19] Actually, Postal's claim is correct only if we consider data where the prenominal modifier is taken to be the antecedent; if we look at data involving the head noun as antecedent, we make the somewhat unexpected discovery that head nouns of CNs **may** act as antecedents for a variety of outbound anaphora (including the pronouns *one*, *those*, and *that*) and in other cases may themselves be pronouns. (For details, see Levi 1977:330–333.) This contradicts Postal's explicit claim that no anaphor may be contained within a "derivative" form, as well as his claim that no single component of a derivative may serve as antecedent for outbound anaphora. Although these facts raise serious questions about the nature of anaphoric island-hood, I believe that they do not vitiate the relevance of the data in (3.22)–(3.25) to the present argument. I must confess, however, that I myself have no explanation at present for the puzzling discrepancy between the anaphoric island-hood of prenominal modifiers within CNs (as indicated in the examples just cited) and the anaphoric accessibility of their head nouns (as illustrated in Levi 1977).

e. *birds-that-she-likes sanctuary
(< sanctuary FOR birds that she likes)

It is, however, an indisputable fact that a CN, once formed by this rule, may itself be input to a further application of the CN formation process, so that (3.27a) may itself become the preposed modifier in (3.27b), (3.27b) in (3.27c), and so forth.

(3.27)   a. [bird] sanctuary
          b. [bird sanctuary] proposal
          c. [bird sanctuary proposal] cover
          d. [bird sanctuary proposal cover] designer

If CNs were not dominated by a node label of N, it would be difficult indeed to characterize in a straightforward and unitary fashion the input to the relevant transformations. (For details of these rules, see Chapters 4 and 5.) However, if we do accept CNs as being dominated by N nodes, we at once simplify the description of CN formation (in what is, moreover, an intuitively satisfying way) and increase our predictive accuracy. This theoretical gain thus constitutes one more reason for analyzing CNs as syntactic nouns.

4

DERIVATIONS OF COMPLEX NOMINALS BY PREDICATE DELETION

4.1 Recoverably Deletable Predicates

A careful examination of the semantic relationships between head nouns and prenominal modifiers in CNs reveals not only that these relationships are not "endless in number" as Jespersen (1942) and others have asserted, but that the variety of these relationships is in fact confined within a very limited range of possibilities. In this chapter, a derivation of CNs by predicate **deletion** will be proposed to account for the larger part of these possible semantic relationships; in the following chapter, a second derivational path involving predicate **nominalization** will be proposed to account for the remaining semantic possibilities. The two derivations correspond to the distinction observable in surface structure between a CN whose two elements are derived from the two arguments of an underlying predicate, as in *apple pie* or *malarial mosquitoes*, and a CN whose head noun is a nominalization of the underlying predicate, as in *city planner* or *presidential refusal*.

4.1.1 Set Membership

The fundamental claim of this chapter is that the larger part of the semantic relationships that may be associated grammatically with the surface structures of CNs can be expressed by a small set of specifiable

predicates that are recoverably deletable in the process of CN formation.
This set is made up of nine predicates: CAUSE, HAVE, MAKE, USE, BE, IN,
FOR, FROM, and ABOUT. These predicates, and only these predicates, may
be deleted in the process of transforming an underlying relative clause
construction into the typically ambiguous surface configuration of the
CN. Examples of CNs derived by deletion of these nine predicates are
given in (4.1); the abbreviation **RDP** in the first column (and henceforth)
stands for Recoverably Deletable Predicate.

(4.1)

RDP	$N_1 <$ direct object of relative clause	$N_1 <$ subject of relative clause
CAUSE	*tear gas* *disease germ* *malarial mosquitoes* *traumatic event* *mortal blow*	*drug deaths* *birth pains* *nicotine fit* *viral infection* *thermal stress*
HAVE	*picture book* *apple cake* *gunboat* *musical comedy* *industrial area*	*government land* *lemon peel* *student power* *reptilian scales* *feminine intuition*
MAKE	*honeybee* *silkworm* *musical clock* *sebaceous glands* *songbird*	*daisy chains* *snowball* *consonantal patterns* *molecular chains* *stellar configurations*
USE	*voice vote* *steam iron* *manual labor* *solar generator* *vehicular transportation*	————
BE	*soldier ant* *target structure* *professorial friends* *consonantal segment* *mammalian vertebrates*	————
IN	*field mouse* *morning prayers* *marine life* *marital sex* *autumnal rains*	————
FOR	*horse doctor* *arms budget* *avian sanctuary* *aldermanic salaries* *nasal mist*	————

FROM	*olive oil*	————
	test-tube baby	
	apple seed	
	rural visitors	
	solar energy	
ABOUT	*tax law*	————
	price war	
	abortion vote	
	criminal policy	
	linguistic lecture	

More traditional terms for the relationships expressed by the RDPs are indicated in (4.2). Many of the categories have figured in analyses of English compounds by other investigators (such as Koziol 1937, Jespersen 1942, Hatcher 1960, Brekle 1970, and Adams 1973), as well as in studies of compounds in other languages (such as Li 1971 on Chinese, or Motsch 1970 on German), but I believe that the present classification constitutes a significantly less redundant and descriptively more adequate account of the observable semantic distributions than any previous treatment of the subject.

(4.2)

RDP	Traditional term
CAUSE	causative
HAVE	possessive/dative
MAKE	productive; constitutive, compositional
USE	instrumental
BE	essive/appositional
IN	locative [spatial or temporal]
FOR	purposive/benefactive
FROM	source/ablative
ABOUT	topic

The division of the examples in (4.1) into two columns reflects the fact that the prenominal modifiers in these CNs can be derived from either the subject or the direct object of the underlying predicate. Thus, *malarial mosquitoes* is derived from an intermediate *mosquitoes which cause malaria*, but *viral infection* is derived not from *infection which causes viruses* but rather from *infection which viruses cause*.[1]

It is interesting to note that this syntactic ambiguity between a deriva-

[1] This statement is true for the sense in which we normally use the expression *viral infection*. Theoretically, however, either intermediate structure (*infection which viruses cause* or *infection which causes viruses*) could underlie this CN. In normal discourse, we disambiguate such forms by the aid of lexical, semantic, and pragmatic clues (to be discussed in Section 4.4).

tion of the modifier from the underlying subject or from the underlying object is possible for only three predicates (CAUSE, HAVE, and MAKE), but does not occur for any of the remaining predicates, as can be seen by the spaces left in the right-hand column of (4.1). The explanation for this division follows from the fact that the CNs which derive their prenominal modifiers from underlying subjects must go through a stage in which the Passive Transformation applies to the S of the relative clause. For example, to derive *viral infection*, the underlying relative clause must be passivized to produce the sequence of intermediate stages sketched informally in (4.3).

(4.3)   a. *infection ## virus cause infection ##*
         b. *infection ## infection is caused by virus ##*   (Passive)
         c. *infection ## infection is virus-caused ##*
                 (Compound Adjective Formation)
         d. *infection ## which ## is virus-caused ## ##*
                 (Relative Clause Formation)
         e. *infection virus-caused*   (WH-*be* Deletion)
         f. *virus-caused infection*   (Predicate Preposing)
         g. *virus infection*   (RDP Deletion: CAUSE$_2$)
         h. *viral infection*   (Morphological Adjectivalization)

A more detailed account of this derivation, including an explanation of the various transformations involved, will be provided in the following section. For the present, however, it suffices to note that the application of the Passive Transformation to produce the structure of (4.3b) is a prerequisite for forming the compound adjective *virus-caused*, from which the nonpred adj *viral* will subsequently be derived.

We can now understand why it is that CNs derived by deletion of the predicates BE, IN, FOR, FROM, and ABOUT do not appear in the right-hand column of (4.1). BE is always blocked from passivization, presumably on semantic grounds (hence, *Suzanne is the mother of Jacques* may not become **the mother of Jacques is been* [=*be* + -*ed*]*by Suzanne*), while the lexical descendants of FOR, IN, FROM, and ABOUT are morphologically blocked from passivization by virtue of their non-verbal status as prepositions.[2]

---

[2] An interesting question is whether these semantic relations would be passivizable in languages where they might surface in overtly verbal, rather than prepositional, form. (Consider, for example, the relation of the English phrases *be located/situated/found at* to the RDP IN, *be intended for* to FOR, *derive/be derived from* to FROM, and *concern/be concerned with* to ABOUT.) Preliminary evidence suggests that passivization must still be blocked on semantic grounds, but this question will not be pursued here.

The case of USE, however, is not so easy to explain. To begin with, it is important to note that the verb *use* that may occur in paraphrases of instrumental CNs (e.g., *clock using electricity* for *electrical clock*) is not the same as the *use* that occurs in expressions like *vehicles used by corporations* or *books used by professors*. The former carries a basically instrumental sense which will be indicated here by use_i, whereas the latter carries a related but distinct agentive sense, which will be shown as use_a.

That the verbs designated here as use_i and use_a actually represent two distinct lexical items is suggested by the different possibilities for CN formation shown in (4.4), as well as by the paraphrase possibilities shown in (4.5).

(4.4)   a. use_i: *clock using electricity* $\Rightarrow$ *electrical clock*
            use_a: *villagers using electricity* $\not\Rightarrow$ **electrical villagers*

       b. use_i: *communication using telephones* $\Rightarrow$ *telephonic communication*
            use_a: *secretaries using telephones* $\not\Rightarrow$ **telephonic secretaries*

       c. use_i: *transport using automobiles* $\Rightarrow$ *automotive transport*
            use_a: *executives using automobiles* $\not\Rightarrow$ **automotive executives*

(4.5)   a. use_i: *communication using phones* = *communication by means of phones*
            use_a: *secretaries using phones* $\neq$ **secretaries by means of phones*

       b. use_i: *communication using phones* $\neq$ **communication putting phones to use*
            use_a: *secretaries using phones* = *secretaries putting phones to use*

The data in (4.4) provide evidence for the claim that use_i and use_a are two distinct lexical items, since CN formation permits the deletion of only the verb with the true instrumental sense (i.e., use_i) but not the agentive verb use_a. The data in (4.5) give further support to this distinction, in that only those expressions analyzed here as containing use_i may be consistently paraphrased by the unambiguously instrumental expression *by means of*, whereas only those expressions analyzed as containing use_a may be consistently paraphrased by the volitional expression *put to use*; the two periphrastic expressions thus mirror the complementary distribution displayed by use_i and use_a.

This particular distinction correlates naturally with the kinds of subject NPs that use_i and use_a normally take. The former generally appears with nominalizations (both transparent and opaque) of activity verbs, such as *communication*, *transport*, or *cure*, which describe the activity that some agent is doing **by means of** the instrumental noun; the agent, however, is not specified in CNs like *vehicular transport* or *telephonic communication*. In contrast, the second sense, *use*$_a$, typically takes animate agents as subject NPs, as in *secretaries using telephones*, *villagers using electricity*, or *executives using automobiles*. In these expressions, the agent and the instrument are specified, but not the activity (although phrases like *for communication*, *for illumination* could of course be added); this sense can be paraphrased by the volitional *put to use* precisely because an agent, rather than an activity, is the grammatical subject.

Although the meanings of *use*$_i$ and *use*$_a$ are certainly related, the syntactic and semantic evidence of (4.4) and (4.5) shows that they are not identical. Of the two meanings, only the instrumental one is encoded in the predicate USE in this study; as a consequence, only lexical items that have been inserted to replace USE (i.e., only tokens of *use*$_i$) are recoverably deletable, and hence relevant here. However, having distinguished the verb *use*$_i$ from the related "agentive" verb, I must still explain the absence of CNs derived from a passivized *use*$_i$. The kind of CN that would fill the empty slot in (4.1) is a CN like *mechanical electricity* which would be derived from an intermediate phrase like *electricity used by machines*; however, I have come across no such examples in 5 years of research. Finding such a phrase is complicated by the fact that superficially good candidates, such as *industrial electricity* from a presumed source of *electricity used by industry*, are more accurately analyzed as deriving from *electricity for industry* by deletion of the well-motivated predicate FOR. Although isolation of the instrumental sense of *use*$_i$ is a necessary step in analyzing the gap in the right-hand column of (4.1), at the present I simply do not have an adequate explanation for the absence of CNs derived from a passivized *use*$_i$.

4.1.2 General Observations

The details of the derivations of the CNs shown in (4.1) will be presented in Section 4.2. For the present, let us observe that the derivations divide into three main types, depending on the morphological characteristics of the compound adjective which is found in all the derivations. The three adjectival types are illustrated by *virus-caused*, *music-making*, and *in-margin*, which represent compound adjectives formed by the combination of a noun with a passive participle, an active participle, and a preposition, respectively. A sketch of the derivational type involving a passive participle has already been given in (4.3); comparable sketches for the active participle and prepositional types are provided in (4.6) and (4.7), respectively.

(4.6)   a. *clock ## clock make music ##*
         b. *clock ## clock music-making ##* (Compound Adjective
            Formation)
         c. *clock ## clock be music-making ##* (Copula Insertion)
         d. *clock ## which ## be music-making ## ##* (Relative
            Clause Formation)
         e. *clock music-making*   (WH-*be* Deletion)
         f. *music-making clock*   (Predicate Preposing)
         g. *music clock*   (RDP Deletion: MAKE$_1$)
         h. *musical clock*   (Morphological Adjectivalization)

(4.7)   a. *note ## note in margin ##*
         b. *note ## note in-margin ##* (Compound Adjective Forma-
            tion)
         c. *note ## note be in-margin ##* (Copula Insertion)
         d. *note ## which ## be in-margin ## ##* (Relative Clause
            Formation)
         e. *note in-margin*   (WH-*be* Deletion)
         f. *in-margin note*   (Predicate Preposing)
         g. *margin note*   (RDP Deletion: IN)
         h. *marginal note*   (Morphological Adjectivalization)

Except for the difference in morphological structure of the compound
adjectives formed in Step (c) of (4.3), and Step (b) of (4.6) and (4.7), the
intermediate stages of the three derivations are almost wholly comparable, including the fact that each derivation shows an application of
Recoverable Predicate Deletion (or RDP Deletion). This rule (or rather,
this set of rules—since there is a separate rule for each RDP) forms a CN
by deleting the predicate element of the compound adjective and
Chomsky-adjoining the remaining noun to the head noun of the construction; for this reason, the RDP Deletion rules might also be called
Complex Nominal Formation rules. Note, however, that of the innumerable predicates that appear in English compound adjectives, only those
which lexicalize a member of the RDP set are deletable by these rules.

   In references made in this study to the nine specific Recoverably
Deletable Predicates, the RDPs will normally appear in small capital
letters in order to stress that it is the semantic structure that is most
relevant to deletability, rather than individual lexical items. For example,
IN represents a general locative predicate, which provides no more meaning than the fact that 'x is located at y'; this location may be in time or in
space, in concrete or abstract places. As it happens, this underlying
predicate may be lexicalized in English by a variety of surface preposi-

tions, including *in*, *on*, and *at*, the selection of which is determined by both semantic and idiosyncratic (lexical) constraints. The relevant Recoverable Predicate Deletion rule, therefore, does not delete the surface lexical item *in* specifically, but rather any node labelled Preposition that corresponds to (i.e., is a lexicalization of) an instance of the generalized predicate IN.

Thus, *terrestrial life* is derived from the same LS that may also surface as *life on earth* (but not as **life in earth*), and *polar climate* shares its LS with *climate at a pole* (but not with **climate in a pole*). The fact that English may lexicalize IN in at least three ways (unlike Hebrew, for example, where the underlying IN is expressed much more consistently by a single preposition) is a superficial phenomenon which does not affect the application of the appropriate RDP Deletion rule; even though specific lexical prepositions have already been inserted by the time that rule applies, the RDP Deletion Transformation must be able to "look back" to the semantic structure—by means of a global rule—in order to see which prepositions are direct descendants of IN, and which are not.[3]

Similarly, we have seen that not all occurrences of the English word *use* will be deletable in CN formation, but rather only those which are in fact descended from the instrumental predicate written here as USE. Both these examples show that the syntactic process of CN formation is inextricably linked with (and, in fact, triggered by) the semantic structures underlying the lexical verbs or prepositions that have been inserted into the IS trees.

The reader will surely have noticed that almost all of the predicates in the RDP set are, in at least a pretheoretical sense, semantically very basic predicates; moreover, a number of these predicates are extremely vague and hence very broad in application. The selection of these highly general predicates can nevertheless be justified on three major grounds. First, the identification of the RDP set which they constitute allows us to make highly accurate predictions about which semantic structures can underlie CNs and which can not; second, the specification of general predicates serves to include a number of more specific relations that might otherwise have to be enumerated individually with a concomitant

---

[3] It is actually more difficult than indicated here to determine the exact point at which lexical insertion operates to replace semantic predicates with specific English lexical items. Although there exists some important evidence to support the approach chosen here (namely, one in terms of **early** lexical insertion combined with a global rule for "remembering" the semantic content), the derivations would work even more smoothly were it the case that lexical insertion occurred relatively late. In the latter case, the predicates might still be present when RDP Deletion applied, so that a global rule would not be at all necessary.

     4 CN DERIVATIONS BY PREDICATE DELETION

loss of generality; third, this analysis creates no more confusion or vagueness than is observable in periphrastic equivalents with overt lexicalizations of the RDPs (e.g., *x has y, x is in y, x is for y*), where **some** fundamental distinctions are made but where a great deal is typically left unspecified.

The value of including more specific predicates within a more general one is perhaps best illustrated by comparing an analysis like the present one, which contains a highly general locative predicate IN, with an analysis which would include more specific relationships than are here subsumed under IN. It would be possible, for instance, to isolate the following relationships and to claim that each defines a separate group of CNs: INHABIT, GROW-IN, ACCORDING-TO, DURING, FOUND-IN, OCCUR-IN. Forms that correspond to each of these relations are indicated in (4.8).

(4.8) a. INHABIT: *desert rat, field mouse, water bugs, city folk, polar bear, aquatic mammals, urban families*

     b. GROW-IN: *water lilies, mountain laurel, tropical lily, suburban crabgrass, sylvan flora, desert blossoms, field weed*

     c. ACCORDING-TO: *logical impossibility, theological fallacy, communist tenet, ideological truth*

     d. DURING: *spring showers, autumnal rains, night flight, nocturnal activity, morning prayers, matutinal devotions*

     e. FOUND-IN: *urban parks, city libraries, hospital wards, textual errors, marginal note, phonological pattern*

     f. OCCUR-IN: *marital adjustments, rural unrest, urban riots, academic transfers, industrial strikes*

The positing of six or more separate groups and as many separate predicates, however, would obscure the generalization that is captured in the present theory by positing a single predicate IN for both spatial and temporal, both concrete and abstract location.[4] By choosing the latter alternative, we can predict with a single statement the general meaning (and well-formedness) not only of *desert rat*, but also of *desert tribes*, *desert flowers, desert winds, desert springs, desert warmth*, and many

---

[4] The choice of capital letters, of course, could just as well be AT or LOC or LOCTEMP, as IN, since any choice is merely a notational device for referring to a particular semantic concept which is presumed to be unitary. For a fuller discussion of the semantic unity of temporal and spatial location, see Traugott 1975.

more. Although it is certainly true that each of these CNs, as it is normally used, might be paraphrased with a far more specific predicate than the general IN (e.g., *living in, growing in, blowing in, flowing in,* or *felt in* for the CNs just cited), we must recognize that (*a*) it cannot be accidental that all of these paraphrases must in any case include the locative preposition *in;* and (*b*) the more specific verbs are wholly predictable on the basis of the semantic characteristics of the surface components of the CN. For example, almost any CN whose head noun denotes an animate creature and whose modifier is a place noun will be a suitable member for the INHABIT group of (4.8a), and thus may have a paraphrase containing a verb like *inhabit* or *live in;* similarly, CNs whose head noun is a nominalization and whose modifier represents a place will generally be interpretable along the lines of the OCCUR-IN group in (4.8f). Since, however, these differences not only are predictable on independent grounds but also seem to have no effect whatsoever on the formation of the corresponding CNs, our description need not include the degree of specificity represented by the groupings in (4.8). Instead, a single predicate IN will be used to generate (and/or interpret) all these CNs without impairing either the descriptive adequacy or predictive power of the theory.[5]

We must also recognize, however, that this goal of reducing redundancy and overspecificity in favor of larger generalizations cannot be pursued beyond a reasonable and empirically motivated point; specifically, we would gain little (and lose much) by replacing the RDPs identified here with an even more general analysis in terms of some virtually unrestricted relationship of 'x is related to y'. Although this approach has been adopted by a number of linguists in the past (e.g., Jespersen 1942 for English, Reif 1968 for Hebrew), it is in fact both descriptively inadequate and predictively useless. To be sure, the RDP analysis proposed here does not (and cannot) specify **all** that we know about the meanings of individual CNs, since parts of the meaning are predictable on independent semantic grounds while other parts are the result of lexicalization and hence inherently unpredictable. Nonetheless, it is claimed here that the RDP analysis captures all the **productive** aspects of CN formation and thus, all the information that a grammar can in any case be expected to predict. In contrast, a description which claims that the only semantic regularity observable in CNs of a N_1–N_2 form is that of 'N_2 relates (is related) to N_1', is implicitly claiming that the

<hr>

[5] Reif's 1968 treatment of Hebrew CNs is a good example of an even more highly overspecified semantic analysis than that sketched here; for more detailed criticism, see Section 8 of Levi 1976.

   4 CN DERIVATIONS BY PREDICATE DELETION

CNs shown in (4.9) are understood in terms of the paraphrases listed there, a claim which has no plausibility at all.

(4.9) *student power*   'power relating/related to students'
         musical clock   'clock relating/related to music'
         women professors   'professors relating/related to women'
         tear gas   'gas relating/related to tears'
         stone walls   'walls relating/related to stone(s)'
         faculty decisions   'decisions relating/related to faculty'

A final point to be made in favor of the degree of specificity incorporated in the RDP analysis presented here is that the rather broad scope and high productivity that characterize most of these predicates are systematically paralleled in periphrastic equivalents of CNs, whether these be in the form of full sentences or reduced relative clauses. For example, the scope of CNs included here under the predicate IN is no larger than the scope of paired NPs that may be included in corresponding full sentences on the pattern of 'x is in/at/on y' or in reduced relative clauses of the form 'x in/at/on y', since the same set of prepositions (*in, on, at*) is regularly used in English to express both spatial and temporal, both concrete and abstract locations. Thus, the impressive variety of CNs formed by deletion of IN is precisely matched by the variety of periphrastic expressions (such as *flowers in the desert, gardens on roofs, sex in marriage, experiences at birth, fallacies in logic,* and *conflicts in academia*) which we use with comparable frequency and comparable interpretability. For much the same reason that we do not need separate lexical entries to express, say, an abstract spatial IN as opposed to a concrete spatial IN, so we also do not need separate predicates in order to generate the different CN groupings shown earlier in (4.8). While it is both interesting and profitable for linguists to examine the many ways in which any one of these predicates may be used (e.g., the many semantic relationships represented by such predicates as HAVE, BE, MAKE, and CAUSE, which may or may not be expressed by English *have, be, make* and *cause*), still it is claimed here that such distinctions are neither necessary nor relevant for that part of the grammar which is to describe the formation of CNs. The present approach, then, is offered as an analysis which incorporates neither the maximal nor the minimal degree of generalization possible but rather an optimal degree. It is hoped that the remainder of this work will justify this claim.

The last general observation to be made in this section is that the members of the RDP set exhibit different frequencies of occurrence; some are immensely productive (both in the sense of having the largest number of attested examples, and in the sense of having the greatest apparent potential for exploitation in creating new CNs), whereas others are so restricted as to raise serious questions about the validity of bestowing membership in the RDP set upon them. In the English data, the least productive predicates are HAVE derived from an active participle (paraphrasable by N_2 *with* N_1, and exemplified earlier in [4.1] by the CNs *picture book* and *apple cake*), CAUSE, MAKE, and FROM; moderately productive are USE, BE, and ABOUT; and most productive are FOR, IN, and HAVE derived from the passive participle form (paraphrasable by possessive *of*, and exemplified in [4.1] by *government land* and *lemon peel*). (The nominalization processes analyzed in the next chapter are also immensely productive in forming CNs.) The data also reveal a systematic but puzzling skew in favor of CN forms based on passivized verbs rather than active ones, so that the deletion of the verbal predicates CAUSE, HAVE, and MAKE produces many more forms that belong in the right-hand column of (4.1) than those suitable for the left; why this is so remains a mystery to me.

Since a parallel study which I carried out on CNs in Modern Hebrew uncovered surprisingly similar distributions (Levi 1976, Section 4), two important questions for future research suggest themselves: How universal are the degrees of productivity for each CN type, and—if comparable frequencies are found across unrelated languages—to what can we attribute these similarities? (Although no attempt to answer these questions can be undertaken here, the issue of cognitive and perceptual universals which they raise is one which should intrigue philosophers and psychologists as well as linguists.)

4.1.3 Specific Predicates

Now that the general characteristics of the RDP set have been presented, we may turn our attention to those aspects of individual RDPs which are pertinent to this analysis. Certain features of the predicates USE and IN have already been discussed; in this section, characteristics of the other RDPs will be analyzed, together with some of the questions that each raises for the overall theory. Examples of each RDP type are provided in (4.1), in the lengthier lists of the Appendix (q.v.), and in certain cases, within the discussion to follow.

One notational convention will be introduced here. To discuss the separate derivational paths involving compound adjectives based on active participles as opposed to those based on passive participles, two subscripts will be used on the RDP in question; for example, CNs listed under CAUSE$_1$ are those derived from a compound adjective with *-causing* (e.g., *malaria-causing mosquitoes*), while CNs associated with CAUSE$_2$ are those derived via a compound adjective in *-caused* (e.g., *virus-caused* *disease*). This distinction corresponds to the left- and right-hand columns, respectively, of the table shown in (4.1). (Viewed in terms of the source structures of these CNs, the two columns in turn correspond to CNs whose modifier is derived from the direct object of the relative clause, as opposed to those whose modifier is derived from the subject of the relative clause.)

CAUSE

The general predicate written here as CAUSE shares with several other RDPs (i.e., MAKE, HAVE, and BE) the characteristic that it may be analyzed as covering several different senses, including direct and indirect causation or "manipulative" and "directive" causation (Shibatani 1976:31–38). Recent treatments of this matter (such as Shibatani 1973, McCawley 1976c, and several articles in Shibatani 1976) have identified a number of grammatical phenomena whose explanations appear to depend on these distinctions. No evidence has yet appeared, however, to suggest that the distinctions analyzed in these works affect in any way the derivations of CNs which express causative relationships; it is for this reason that CAUSE is treated as a single predicate here.

We may note, nonetheless, that the current investigations into grammatically significant distinctions among causative expressions of various types may eventually help to explain some of the problems connected with the CAUSE data in this study. For instance, I have no explanation for the fact that the number of CNs in my data that suggest a CAUSE$_1$ derivation (i.e., via an intermediate stage containing a compound adjective of the form N-*causing* rather than N-*caused*; cf. the two columns of data in [4.1]) is extremely small; in fact, *malarial mosquitoes, traumatic event, tear gas,* and *disease germ* comprise almost half the corpus. (Interestingly, a correspondingly small proportion among the CAUSE$_1$ CNs in the Hebrew data analyzed in Levi 1976 indicates this may not be an English-specific quirk.) A related problem is that the apparently comparable expressions, *cancer-causing chemicals* and *pollution-causing processes,* may not undergo predicate deletion to become *cancer chemicals* or *pollution processes.* I can suggest no reason for this latter fact at present.

The two problems just mentioned, however, seem to suggest that the input to CAUSE Deletion may be systematically restricted, for some as yet undiscovered reason, to compound adjectives formed in *-caused* rather than *-causing.* (This would be the reverse of the restriction which appears to hold for CNs derived from the predicate USE, discussed earlier; in that case, CNs are formed only if the intermediate stage contains a compound adjective ending in an active participle rather than a passive one.) In such an event, an alternative analysis would have to be found to account for the CNs that are here derived from an 'N-*causing*' intermediate form. (For a full list of such forms, see the Appendix under CAUSE₁.)

HAVE

This particular predicate is of interest since the participial forms that we would expect in the intermediate derivational stages never surface per se, but rather are regularly replaced by suppletive prepositional phrases (which then may or may not have the preposition deleted to form a CN). Thus, the underlying form *cake that has apples* might surface either as *cake with apples* or as the CN *apple cake*, but never as **cake that is apple-having* or **apple-having cake*; similarly, the form *peel that a lemon has* could be transformed into *peel of a lemon* or the CN *lemon peel* but never into **peel that is lemon-had* or **lemon-had peel*. Why the participial forms of this particular verb should be blocked from forming any compound adjectives is not clear, although this certainly is not the only idiosyncratic aspect of this most basic, multipurpose verb.

On the other hand, if we recall that many of the world's languages (including Proto-Indo-European) do not have a morphological verb corresponding to English *have*, but use prepositional phrases and/or case markers to express the comparable relationships, then the replacement in English of participial *have-* forms by phrases with possessive *of* and *with* begins to seem somewhat less strange as well as somewhat less arbitrary.[6]

It was noted previously that the HAVE₂ possessive relation is one of the most productive of all the RDP categories (although HAVE₁ is among the least productive). At least a partial explanation of the freedom with which CNs based on HAVE₂ are formed may be found in the observations of Bach (1967) on two of the vaguest—and most useful—verbs in the English language, *have* and *be*:

> The two forms are distinguished syntactically from most true verbs by the fact that they have no selectional restrictions in themselves, but occur in constructions where the selections reach across from subject to 'object' or complement. Likewise, from a semantic point of view, their contribution to the meaning of a sentence is determined completely by the items that they link [pp. 476–477].

Thus, unlike IN (which requires a locative noun for its object), or CAUSE (for which pragmatic considerations severely constrain the number of NP pairs that it can appropriately link), the predicate HAVE enjoys considerably greater play across the lexicon in selecting nominal arguments; it

---

[6] Indeed, in view of the fact that the RDP set includes a high proportion of relationships that are apparently well suited to lexicalization in the form of either prepositions or case markers, perhaps the question that should be addressed here is why are not **all** the RDP relationships equally well transformable into these syntactic forms?

should therefore come as no surprise that the list of CNs that could be generated under HAVE$_2$ is virtually endless.

The fact that HAVE$_1$ CNs do not appear with anywhere near the same frequency may be due in part to the fact that although the CN construction is well suited to the function of identifying a possessed object or characteristic by naming both it and its "possessor" (e.g., *student power, horsepower, candlepower, presidential power, national power*), the comparable task of picking out a "possessor" by naming something distinctive that it possesses is much more frequently accomplished by means of denominal adjectives that have **incorporated** within them the meaning of HAVE$_1$. Thus, a person distinguished by the wealth he or she possesses would be referred to as a *wealthy* [= 'having wealth'] *person*, rather than as a *wealth person*; similarly, a village characterized by harmony is more likely to be called a *harmonious* [= 'having harmony'] *village* than a *harmony village*. (Interestingly, both HAVE$_1$ and HAVE$_2$ CNs are often paraphrasable by genitival equivalents, as in *person of wealth, village of harmony, power of students, students' power*; this is not surprising, however, in view of the fact that the genitive shares with the CN construction both a highly functional expressive conciseness and a concomitant ambiguity on both syntactic and semantic planes.)[7]

MAKE

The CNs derived by the deletion of MAKE$_1$ (i.e., by the deletion of an active participle *-making*) are shown in (4.10), while some of those derived by the deletion of MAKE$_2$ (i.e., by the deletion of a passive participle *-made*) are shown in (4.11).

(4.10)

$$\begin{matrix} silkworm \\ honeybee \\ songbird \\ sap\ tree \end{matrix} \left\{ \begin{matrix} sweat \\ salivary \\ sebaceous \end{matrix} \right\} glands \quad \begin{matrix} music\ box \\ musical\ clock \end{matrix}$$

(4.11)   a. *daisy chains*
           molecular chains
           consonantal patterns
           floral wreath
           root system

---

[7] As a consequence, the genitive is the syntactic option most heavily exploited for adjoining two nouns by those languages which are poor in denominal adjectives and/or do not permit the formation of N–N CNs; the first case is exemplified by Akkadian and the second by Icelandic. (These genitival exploiters were pointed out to me by Gene Gragg and Jerrold Sadock, respectively.)

> b. *landmass*
>   *chocolate bar*
>   *stone wall*
>   *sugar cube*
>   *bronze statue*
>
> c. *warrior caste*
>   *immigrant minority*
>   *worker teams*
>   *student committee*
>   *youth corps*

It appears that the two groups, originally distinguished on the basis of participial type within the compound adjective, are systematically distinguishable on additional, semantic grounds. The forms in (4.10) are derived from underlying forms in which the predicate MAKE seems to have a sense of 'physically producing, causing to come into existence'. This sense is not at all present in the CNs of (4.11), which are apparently derivable from sources where MAKE has a compositional reading (i.e., 'x made up of y, x made out of y'). The three groups shown in (4.11) share the latter relationship, but are distinguishable on independent semantic grounds. Specifically, the CNs in the (a) group each contain a modifier denoting a unit, and a head noun denoting a configuration; the CNs in the (b) group have head nouns that denote either a mass or an artifact of some sort, and modifiers that describe its constituent material; and the CNs in the (c) group have head nouns describing human collectives, and modifiers which specify their membership. (It is important to note that the CNs in the middle group all have an alternative analysis under BE, since we can just as readily say that *chocolate bar* derives from *bar which is chocolate* as from *bar which is made of chocolate*; this analytic indeterminacy, which is not to be confused with real semantic ambiguity in the surface form, will be discussed in more detail in Section 7.3.) Still, this difference in the pairs of associated nominals does not change the fact that all three groups in (4.11) may be analyzed as sharing the same underlying predicate.

It is, however, much less certain whether the predicate identified thus far as MAKE$_1$ is indeed identical to that shown here as MAKE$_2$. For one thing, the productive and compositional senses expressed at different times by *make* are not as intimately related as, say, the two senses of *use*$_i$ and *use*$_a$ discussed earlier. Moreover, the fact that in 5 years of research, the only examples of MAKE$_1$ CNs that I have found are those shown in (4.10) does suggest that a rule for forming CNs by deletion of this predicate in its active participle form may be insufficiently motivated. This conjecture is further supported by the fact that several of the forms in question (*musical clock, music box, sweat/sebaceous/salivary glands*) could just as well be derived by FOR Deletion as by the MAKE$_1$ Deletion proposed here.

On the other hand, one feature of the present analysis is that we cannot always determine with certainty a **unique** derivation for a given CN; for example, it is not immediately obvious whether *suspense film* should be derived from *film that causes suspense* or *film that has suspense*, or whether *job tension* is best derived via *tension caused by the job*, *tension that the job has*, or *tension on the job*.[8] As a consequence, the fact that some of the forms in (4.10) have alternative analyses is not in itself proof that one of the analyses must be right and any others wrong.

Another reason to group the two sets of data together is that it seems unlikely to be an accident of the English lexicon that both sets of CNs can be paraphrased with expressions using the same lexical verb (*make*). In addition, an argument could be advanced to the effect that the substitution of separate predicates such as PRODUCE (for MAKE$_1$) and FORM or CONSTITUTE (for MAKE$_2$) would result in the loss from the RDP set of a more general predicate in favor of two predicates with significantly more restricted productivity. Since most of the RDPs are semantically extremely broad, the substitution of two narrower members for one broader one should not be made without good reason; such a substitution might well be comparable to the trading of generality for redundancy that would be entailed if the general locative IN were to be replaced by the much more specific predicates listed earlier in (4.8).

For the present, then, the predicate MAKE will be retained as a member of the RDP set which enters into CN formation in both active- and passive-participle forms. It is clear, however, that the strength of the evidence in favor of this analysis is hardly overwhelming.

USE

The major features of this instrumental predicate have been discussed in the preceding section. One additional point to be made is that for those CNs which are derived by USE Deletion and whose head noun denotes an

---

[8] As already noted, this problem is addressed more fully in Section 7.3. For the present, we may simply note that part of the problem stems from the fact that knowledge of the **referent** of a CN like *job tension* or *suspense film* is not equivalent to knowledge of, or insight into, the grammatical derivation appropriate to such a form. In addition, speakers who use a given form may well have different mental images associated with it, so that assigning a **single** underlying source may make sense for some of these CNs only at the idiolectal level.

object rather than an activity, the most acceptable paraphrases will make explicit the notion of 'x using y in order to function'; for example, *steam iron* might be paraphrased as *iron using steam to function* or *iron functioning by means of steam*, rather than the shorter *iron using steam* or even the ungrammatical **iron by means of steam*. In contrast, paraphrases of USE CNs whose head nouns are nominalized verbs will not need this addition, so that *vehicular transport* or *shock treatment* can be paraphrased as *transport using vehicles* or *treatment by means of shock*. It seems that the optimal expression of an instrumental relation (except within the compressed form of a CN) must include some indication of the activity for which the presupposed agent is using the instrument; it is to indicate this activity, then, that a verb like *function* is felt to be needed.

BE

Although this very basic predicate underlies a number of semantically distinguishable classes, it is claimed here that these divisions are predictable on independent semantic grounds and make no empirical difference in the actual derivation of the CNs in the various subgroups. Three of the major semantic subtypes are the compositional, the "genus–species," and the metaphorical. (4.12) shows CNs of the compositional type, whose modifier denotes the material of which the head noun is composed; all of these have an alternative analysis under MAKE$_2$, as indicated earlier.

(4.12)
snowball	*stone wall*
water drop	*glass cup*
landmass	*bronze statue*
sand dune	*copper coins*

In the "genus–species" subset of BE CNs, illustrated in (4.13), the head noun denotes a category of which the modifier isolates a subtype. These might be real biological subdivisions, as in *pine tree* or *mammalian vertebrates*, or more informal classificatory relationships, as in *sports activities*, *consonantal segments*, or *cash basis*.

(4.13)
pine tree	*winter season*
flounder fish	*sports activities*
cactus plant	*cash basis*
mammalian vertebrates	*murder charge*
collie dog	*consonantal segments*

A third subset of the CNs derived by BE Deletion is composed of those regularly used in a metaphorical sense. Note, however, that no claim is made here that the derivation of these forms is affected in any way by the metaphorical reading normally imposed on them; indeed, I claim that **no** grammatical description can predict either that the CN is in fact a metaphor or the way in which it is so taken (e.g., that the *finger lakes* resemble fingers in both number and shape, whereas *queen bees* and *soldier ants* are so-called primarily because of their function and behavior). The motivation for identifying the CNs in (4.14) as a separate group is thus more didactic than theoretical (i.e., simply to call to the reader's attention the fruitfulness of the predicate BE for deriving both lexicalized and creative CN metaphors).[9]

(4.14)	*mother church*	*frogman*
	sister node	*hermit crab*
	satellite nation	*soldier ant*
	finger lakes	*queen bee*
	wastebasket category	*target structure*

Two other groups of CNs that might also be treated under BE (besides the exocentric examples excluded in Chapter 1) are the "coordinate" type, such as *secretary–treasurer* or *panty–girdle*, and the "reduplicated" type, such as *a house house* or *a college college*; for separate reasons, however, neither group will in fact be included formally in this study. Members of the first group (which also includes examples like *king–emperor, gentleman–farmer, dinner–dance, fighter–bomber, broiler–oven*) resemble CNs in their N–N structure, but are systematically distinguishable from regular CNs by the fact that the surface nouns are semantically coordinated and by the concomitant fact that the surface forms must be viewed as exocentric in nature. Thus, unlike endocentric CNs such as *student listeners* which denotes 'listeners who are students' and thus identifies a subset of the set denoted by the head noun, a coordinate CN like *speaker–listener* in fact means 'person who is [both] a speaker and a listener'; in the latter case, the actual head of the nominal is deleted and only the two predicate nominals of the underlying relative clause reach the surface.

Evidence that the underlying form of the coordinate group actually does include coordinate nominals may be gathered from the fact that the two nominals in this type of expression are always semantically parallel to a degree which is appropriate to well-formed conjunctions but which

[9] Although some linguists would prefer to analyze the CNs in (4.14) as being based on a relation more like LIKE or RESEMBLE, evidence why this approach is inferior to the present one will be presented in Section 4.1.4. In addition, the interested reader is referred to Reddy 1969 for a critique of linguistic analyses of metaphor that are based on violations of context-independent selectional restrictions; Reddy argues convincingly, in fact, that "all statements may be potential metaphors [p. 241]."

is—not surprisingly—not found in other BE-type CNs; thus, *secretary–treasurer* is the conjunction of two job titles, *king–emperor* of two exalted positions, and the more lowly *panty–girdle* of two articles of clothing. In contrast, other BE examples lack this parallelism, as may be seen in forms like *pine tree*, *stone wall*, *hormonal agent*, or *novelty item*. We may note in addition that the exocentric-type paraphrases suitable for forms like *secretary–treasurer* or *panty–girdle* are quite weird when used for the appositional type of BE-derived CNs (e.g., 'something which is both a pine and a tree' for *pine tree*, or 'something which is both stone and a wall' for *stone wall*).[10]

Although one is unlikely ever to find examples of the "reduplicated" type of CN in a formal study, this may be for stylistic rather than theoretical reasons since they are typically ephemeral and colloquial coinages. It remains true, however, that virtually any noun may be duplicated to form what at least looks like a CN, in dialogues like these:

(4.15)　　a. *We've rented a house in Michigan.*
　　　　　　Oh? Is it a hoúse house or just a cabin?

　　　　　b. *Jack blew all his money on a suit.*
　　　　　　Did he buy a súit suit or a leisure suit?

　　　　　c. *I can't figure out what Jerry meant by this question.*
　　　　　　Do you think it's a quéstion question, or just a rhetorical one?

In these examples, the reduplicated forms function in much the same way as qualifying adjectives like *real* or *true* ("Is it a **real** house?" "Was that a **true** question?"), namely, to ask whether the noun was intended to have the standard, traditional, and perhaps most literal meaning of whatever its common definition is thought to be, rather than some nonliteral, peripheral, derivative, or idiosyncratic sense. The problem of determining how speakers and listeners agree on what constitutes the core of a definition (or indeed, whether they really do at all) is a fascinating one, but clearly one which cannot be addressed within the bounds of this study.[11] In any case, reduplicated nominals like those shown in (4.15) have been excluded from this study, partly on the grounds that both their semantic function and their syntactic derivation are not yet well understood, and partly on the grounds that they may in fact be just one special case of a more general linguistic mechanism for pinning down the in-

---

[10] The analyses of compounds in both Lees 1960 and Marchand 1969 also exclude this exocentric coordinate type from consideration within the endocentric appositive type.

[11] Some recent investigations of this problem are reported on in Lakoff 1972, Markowitz 1977, Rips *et al.* 1973, Rosch 1973, and Zadeh 1971.

tended meaning of **any** major constituent. (Note that we can use the same process with verbs and adjectives, as in "Are you going to swim swim, or just paddle around?" or "Are the curtains blúe blue, or more turquoise?") In that case, the duplicating process would have to be described by a broader rule than any of those that form complex nominals.

IN

We turn now to a consideration of IN, the first of the four predicates of the RDP set which regularly surface as prepositions in English. These have been included in the RDP set, along with predicates that are normally lexicalized as verbs, because both kinds of predicates function in essentially the same way: They both take nominal objects, they both form compound adjectives with these objects, and they both may be regularly deleted from these compound adjectives to form complex nominals. In this work, then, the surface prepositions that lexicalize the underlying IN, FOR, FROM, and ABOUT are derived from semantic predicates in exactly the same way that surface verbs, such as *cause* or *produce*, are derived from their antecedent predicates; the different surface category labels that are ultimately assumed by LS predicates must be regarded as reflecting superficial distinctions which disguise their similar origins. In fact, all four of these "prepositional" predicates are occasionally lexicalized in English as surface verbs, namely, as *be located (at)*, *be intended (for)*, *derive/be derived (from)*, and *concern/be concerned (with)* (for IN, FOR, FROM, and ABOUT, respectively), which convey no different or more precise semantic information than the prepositions alone. (For additional arguments in favor of a generalized treatment of prepositions as predicates, see Becker and Arms 1969, and Geis 1970.)

If we now consider IN specifically, we see that its inclusion within the RDP set is justified by the paraphrases and nonparaphrases shown in (4.16)–(4.19), which include a variety of locative prepositions.

(4.16)   a. *friendships in the office = office friendships*
                              = ?in-office friendships
     b. *friendships out of the office ≠ office friendships*
                              = out-of-office friendships

(4.17)   a. *transportation in cities = city/urban transportation*
                             *= *in-city transportation*
     b. *transportation between cities ≠ city/urban transportation*
                             = intercity transportation
                             = interurban transportation

(4.18)   a. *sex in marriage = marital sex*
                    *= *in-marriage sex*

b. *sex outside of marriage ≠ marital sex*
$$= \textit{extramarital sex}$$

(4.19) a. *life on earth = earth/terrestrial life*
$$= \textit{*on-earth life}$$
b. *life under the earth ≠ terrestrial life*
$$= \textit{subterranean life}$$
c. *life* $\left\{\begin{matrix} \textit{not on} \\ \textit{?off} \end{matrix}\right\}$ *the earth ≠ terrestrial life*
$$= \textit{extraterrestrial life}$$

In all these examples, the only prepositional phrases that can be transformed into CNs (by deletion of the preposition from the intermediate compound adjective) are those in which that preposition (here, *in* or *on*) is derived from the simple locative IN. All other prepositions must appear overtly in preposed modifiers, either as Romance prefixes (*sub-* for UNDER, *inter-* for BETWEEN, etc.) or in their independent morphological shape as part of a compound adjective (as in *out-of-office* in [4.16] or the comparable *between-term, under-par, overage, offshore*, etc.). These data thus demonstrate that although many locative prepositional phrases can be transformed into compound adjectives and moved to prenominal position, only those based on the locative predicate IN may have their preposi-. tion deleted from the compound adjective.

Finally, the reader is reminded that this predicate serves to express relations of concrete location (as in *desert mouse, marine life, city parks*), abstract location (*marital sex, logical fallacy, professional specialization*), and temporal location (*night flight, childhood dreams, winter activity*).

FOR

As noted earlier, this "predicate of purpose" is immensely productive in the generation of CNs from intermediate sources of the form 'x is for y'. We must note, however, that the lexical item *for* which may appear in such forms has a variety of sources, only one of which is the predicate of purpose or intent written here as FOR. Other sources for lexical *for* include (*a*) a predicate in the sense of 'favoring' as in *vote for abortion* or *women for peace*; and (*b*) what I claim is a semantically empty object marker for certain verbs as in *appeal for money* or *request for time*. Naturally, in the present context, these other senses will be excluded as irrelevant, since NPs in which *for* means 'favoring' may not form CNs (thus, *vote for abortion* can only become *pro-abortion vote* but not *abortion vote*) and NPs in which *for* is an object marker can prepose the object and delete the preposition according to the derivational steps proposed in Chapter 5 for nominalizations, but not according to the steps proposed in this chapter for predicate deletion.

A sample of CNs derived by FOR Deletion is shown in (4.20).

(4.20)   *bird/avian sanctuary*   from   *sanctuary for birds*
            aldermanic salary   *salary for an alderman*
            industrial equipment   *equipment for industry*
            doghouse   *house for dogs/a dog*
            administrative office   *office for administration*
            cooking/culinary utensils   *utensils for cooking*
            nose drops   *drops for the nose*
            picture album   *album for pictures*

Note that in many of these cases, the paraphrases shown to the right of each CN could be expanded to provide synonymous expressions of the form: [x is for V-ing y], where V stands for some appropriate activity verb. Thus *avian sanctuary* could be paraphrased as *sanctuary for protecting birds*, *aldermanic salary* as *salary for paying to an alderman*, and *administrative office* as *office for handling administration*, although it is significant that these senses are largely predictable from the meanings of the head nouns. In other cases, it appears that the specific verb commonly associated with a given CN of this type must be learned individually, and hence is a datum for the lexicon to **list** rather than for the syntactic component to **generate;** examples of this sort (and plausible associated verbs) include *nose drops (decongest)*, *headache pills (reduce/relieve/stop)*, *picture album (store)*, *automotive shop (repair)*, *clothing shop (sell)*, and *lightning rod (attract/ground)*.

Since, however, the goal of the present analysis is to capture the broadest generalizations possible about the **productive** aspects of CN formation, we must exclude from our description both those generalizations or patterns which are predictable on independent semantic grounds (whose description would therefore duplicate semantic statements made elsewhere in the grammar), and that information which is peculiar to individual CNs and hence appropriate to statement (i.e., listing) in the lexical component. Moreover, even if we could build into our grammatical description the specification of individual verbs for these CNs (ignoring for the moment that in many cases this could be done in no more revealing way than by a simple list), such an approach would fail to reflect the multiple ambiguity that is undeniably characteristic of all CNs; specifying one verb for any particular form would then constitute a misleading and incomplete description of the full range of semantic structures that could be associated, regularly and predictably, with that surface form.

In contrast, the analysis provided here captures the essential generalization that all NPs of the form [N₂ *for* N₁] (again, where *for* < FOR) may be transformed into CNs of the form $[N_1 + N_2]$ regardless of the possible verbs that could be supplied.

Let us illustrate the point with an example taken from Lees 1960. My claim here is that the CN *bull ring* is derivable via an intermediate *ring for bulls* (where *for* < FOR), regardless of what verb might be supplied to indicate more precisely what is to be done with the bulls. Thus, a ring for fighting bulls can be called a *bull ring*, as in the most familiar case; but a ring for simply holding bulls, as in a stockyard, could also be called a *bull ring*, as could a ring for exercising, washing, counting, mating, wrestling, riding, or observing bulls. The generalization to be captured here is that rings intended and/or used for any of these purposes could appropriately be called *rings for bulls*, and that it is from this structure alone that the CN *bull ring* is derived by deletion of the RDP FOR.

This analysis has the additional advantage that CNs like *bull ring*, *bird cage*, *avian sanctuary*, and the many others that can be derived by deleting FOR need not be described as *n*-ways ambiguous, where *n* is the number of possible verbs that might be supplied in more precise definitions.[12] Instead, these CNs will be most correctly described as **vague,** with their vagueness directly attributable to the extreme vagueness of the predicate FOR itself.

In almost all cases, the predicate FOR indicates only this: that a relationship of intent or purpose exists (or is believed to exist) between the objects named by its two underlying arguments. While it is true that such a relationship normally presupposes the existence of a volitional agent to whom the relevant intention or purpose may be ascribed, it is in fact irrelevant to the derivation of CNs derived by FOR Deletion whether (*a*) the relationship expressed by FOR is deliberately established by some creative act of the agent (as in *instructional materials*, used to denote materials created by someone expressly for the purpose of instruction); (*b*) the relationship comes into being through some agent's more or less arbitrary decision (as in CNs like *bingo night*, *basketball season*, and *laundry days*); or (*c*) a natural relationship is simply acknowledged by the speaker without any specific agent being assumed (as in CNs like *digestive*

---

[12] The awkward, and ultimately untenable, position that the multiple ambiguity of compounds is of an order of magnitude comparable to that of some vast number of "possible" transitive verbs is embraced explicitly by Gleitman and Gleitman (1970:94), following the more implicit example of Lees (1960). This position is one of the points most frequently singled out by critics of the latter as allegedly demonstrating the gross inadequacy of **any** transformational derivation of compound forms. However, as Lees himself notes in a later work (Lees 1970:182), this problem falls by the wayside once the strictly syntactic approach typified by Lees 1960 is replaced by an analysis which permits the expression of the appropriate semantic generalizations (represented here by the RDP set).

system and *growth hormone*).[13] The predicate FOR itself (and the preposition *for* that is its normal lexicalization in English) expresses none of these distinctions.

Moreover, FOR conveys no other additional details concerning possible variations on this general theme of "purpose" that might arise in relation to the referents of specific CN forms. Such details can only be supplied or deduced, if at all, from a knowledge of the meaning of the two surface elements, and the relationships that **extralinguistic** knowledge permits us to infer are both plausible and pertinent, in a given context, for the objects denoted by those elements.

It is important to note in this connection that the characteristic vagueness of the predicate FOR permits a speaker to assign to superficially parallel CNs like those in (4.21) two diametrically opposed interpretations, as indicated here:

(4.21)   a. *fertility pills = pills for fertility* [*to increase it*]
             headache pills = pills for headaches [*to decrease them*]

         b. *bug spray = spray for bugs* [*to harm them*]
             pet spray = spray for pets [*to help them*]

         c. *mothballs = balls for moths* [*of something noxious*]
             birdballs = balls for birds [*of something nourishing, like suet*]

However, the semantic oppositions shown in (4.21) are a function not of a structural ambiguity of the predicate FOR, but rather of opposing **assumptions** about the most plausible relationships between the two surface elements; we normally assume that *headache pills* must mean 'pills for suppressing headaches' and *fertility pills* must mean 'pills for enhancing fertility'. My theory claims, however, that all we know for sure is that there is a relationship of intent or purpose between the head noun *pills* and its prenominal modifier (i.e., that the pills are intended to do some **unspecified** thing to, or with, or in connection with, headaches, or fertility, or gout, or whatever). Note, however, that *headache pills* could certainly mean 'pills for inducing or increasing headaches' in the context

---

[13] In a small number of CNs analyzed here under FOR, the relationship appears to be more one of an observed suitability, rather than of deliberate intention on anyone's part; examples of this type include CNs like *beach weather*, *mathematical mind*, *musical talent*, and *swimming build*. These forms also manifest a systematic ambiguity between readings of 'x is for y' and 'x is good for y'. Because of the latter possibility, the unmodified CN may be used synonymously with the same CN preceded by *good*; thus, *beach weather* (without prenominal modifiers) may be used to mean the same thing as *good beach weather*, and *a mathematical mind* to mean the same as *a good mathematical mind*. This is in sharp contrast with most other CNs derived by FOR Deletion for, as many readers will surely recognize, the CN *instructional materials* is in no way synonymous with *good instructional materials*.

of some physiological experiment with that purpose in mind; it would then be just as reasonable, **and** contextually unambiguous, for an experimenter to say (4.22a) to a subject as it would be under normal circumstances for a gynecologist to say (4.22b) to an infertile patient.

(4.22)   a. *Please take these headache pills each morning.*
           b. *Please take these fertility pills each morning.*

As Zwicky and Sadock (1975) have pointed out, proving that a particular lexical item or syntactic configuration is ambiguous is in many cases easier than proving that it is vague. This is due in part to the fact that, in order to demonstrate ambiguity, it suffices to find a single (independently-motivated) transformation that distinguishes between two allegedly distinct readings of a lexical item or phrase. However, the failure to find any transformation that can distinguish between two such allegedly distinct readings (i.e., structurally distinct, rather than semantically unspecified) is not in itself a demonstration of vagueness, since the set of available transformations may **accidentally** fail to distinguish between the two readings of a real ambiguity. (See Zwicky and Sadock 1975 for additional discussion.)

The data in (4.23) to follow therefore do not demonstrate absolutely that FOR is vague rather than ambiguous, but they do nonetheless support such a claim. If *for* (< FOR) were **ambiguous** in the expressions of (4.23) between a positive reading in the sense of 'favoring, enhancing, benefiting' and a negative reading in the sense of 'inhibiting, suppressing, injuring', then we would not expect to be able to apply Conjunction Reduction to nominal forms with these opposing readings. The data here indicate, however, that such Conjunction Reduction is possible (although the reduction somehow sounds more acceptable to me in CNs than in the unreduced prepositional phrases).

(4.23)

a. *She bought a year's supply of pills for* $\begin{Bmatrix} \textit{headache and gout} \\ \textit{?headache and fertility} \end{Bmatrix}$.

   She bought a year's supply of $\begin{Bmatrix} \textit{headache and gout} \\ \textit{headache and fertility} \end{Bmatrix}$ *pills.*

b. *Consumers Union has tested many sprays for* $\begin{Bmatrix} \textit{bugs and roaches} \\ \textit{?bugs and pets} \end{Bmatrix}$.

   Consumers Union has tested many $\begin{Bmatrix} \textit{bug and roach} \\ \textit{bug and pet} \end{Bmatrix}$ *sprays.*

The oddness that some readers may sense in some of the "crossed reading" examples in (4.23) may be due to a psychological set (similar to that described in Zwicky and Sadock 1975:21) which is violated by the juxtaposition of opposing presuppositions in conjunctions; note, however, that if additional information is added to explain the opposition, the reduced phrase appears perfectly normal, as in (4.24):

(4.24)   *Sharon has gotten tremendous headaches from worrying about her inability to conceive, so her doctor sensibly prescribed both headache and fertility pills.*

The preceding discussion provides a useful illustration of one of the pitfalls that must be avoided in trying to determine accurately the set of possible meanings for a given CN, namely, the danger of characterizing either a customary reading, or the first reading that pops into our sometimes slothful and uncreative minds, as the only plausible reading for that form. Our language is a far more complex and versatile instrument than is reflected by such facile analyses, and grammatical descriptions arrived at by totalling up the sum of such snap judgments must necessarily be inadequate. Translating this general caveat into the specific context of CNs derived by FOR Deletion, we must conclude that the association of a specific verb with each of the CNs listed here under FOR would reflect only the nature of the real-world object **customarily** named by that CN, rather than the real range of meanings that our language permits to be expressed by that single form. As a consequence, FOR will continue to be treated here as the single predicate (albeit an extremely vague one) from which the examples in this section and their counterparts in the Appendix are transformationally derived.

FROM

The predicate FROM is used to derive those CNs whose prenominal modifier denotes the **source** of the head noun; in the vast majority of cases, the source is a natural object of either vegetable or animal form (as in *olive oil, cane sugar,* and *wood ash* for the former, and *pork suet* and *alligator leather* for the latter). This predicate is nowhere near as productive as some of the other RDPs already discussed; nonetheless, the number of examples that can be analyzed under FROM (and apparently under no other predicate already in the RDP set) is significant enough to warrant its inclusion among the RDPs.

What is unusual about this predicate, however, is the remarkable degree of semantic homogeneity observable in the CNs it seems to generate; no other category of CNs (with the possible exception of ABOUT, to be discussed next) is quite as uniform in the range of objects that it designates. Thus, almost all examples that fall into this category arrange themselves into just two subgroups: the *olive oil* type shown in (4.25a), and the *country visitor* type shown in (4.25b).[14]

(4.25) a. *olive/castor/mineral oil*   *wood/tobacco ash*
           cane/beet/corn sugar   *wood shavings*
           grain/wood alcohol   *coal dust*
           rye/corn whiskey   *pork suet/bacon grease*
           peanut/apple butter   *alligator leather*

        b. *store clothes/teeth/medicine* *country/rural visitors*
           test-tube baby   *sea breeze*
           kennel puppies   *farm boy(?)*
           country butter

In the examples in (4.25a), the head noun denotes a product or by-product obtained by some kind of processing activity from the natural source named by the modifier; this may involve a natural process (e.g., melting for *bacon grease* or burning for *wood ash*), a technological process (e.g., distillation for *corn whiskey*), or some sort of human manipulation (e.g., crushing for *peanut butter* or planing for *wood shavings*). The forms in (4.25b) are different in that the modifier generally denotes a previous location (and thus is a locative noun); in some cases, the modifier tells us where the head noun is or was obtained (e.g., *store clothes*, *test-tube baby*), whereas in others the modifier simply describes a past habitual location (e.g., *country visitors* in the sense of 'visitors from the country' or *sea breeze*).[15] (Into this group would also fall that large group of CNs that have proper nouns or PPAs as prenominal modifiers denoting 'place of

---

[14] One additional set of examples that might be analyzed under FROM consists of terms for parts of a plant or animal which have become separated from their biological source (e.g., *peach pit*, *apple seed*, *oak leaf*, *duck eggs*, and *goat milk*). Since, however, the same terms are used to denote these objects prior to physical separation or detachment, and since a HAVE$_2$ analysis seems perfectly adequate for these meanings (e.g., *peach pit* from the same source as *pit of a peach* and *pit that a peach has*, or *goat milk* from the same source as *goat's milk* or *milk that a goat has*), listing these forms under FROM would be equivalent to claiming that they are ambiguous between a "pre-separation" meaning (analyzed under HAVE$_2$) and a "post-separation" meaning (analyzed under FROM), a claim which I do not wish to make.

[15] Although the semantic relation expressed in the CNs of (4.25b) might be alternatively analyzed as some combination of PAST plus the general locative IN, such an analysis has not been made here for two reasons. First, no other RDP group appears to be based on a combination of any two predicates (much less a regular RDP plus a tense predicate); secondly, a PAST + IN analysis could not be used to derive the CNs in (4.25a), whereas a FROM analysis is both sufficiently general and semantically adequate to include both groups in (4.25) within a single derivational type.

                    4 CN DERIVATIONS BY PREDICATE DELETION

origin', as in *Florida oranges*, *Israeli leather*, *Swiss cheese*, and *Japanese radios*.)

In all these cases, however, a single derivation via FROM Deletion will be adequate to account for the CNs of (4.25) since there is no evidence that the semantic distinctions observable there have any effect whatsoever on the derivation of these forms. Moreover, these semantic distinctions must in any case be accounted for elsewhere in the grammar and so need not be duplicated in a description of CN formation. (For example, more general semantic statements must be able to characterize the difference involved in identifying as a source some natural object such as *wood* or *grain* as opposed to a large geographical area such as *country* or *sea*; other distinctions might follow from culture-specific, lexicalized associations, as in the case of *test-tube baby* as opposed to *store teeth*.)

ABOUT

The first reason that ABOUT must also be regarded as an unusual RDP is that it has a larger number of lexical equivalents in English than most RDPs; specifically, the semantic relation represented here as ABOUT may be expressed as *concerning* (or *concerned with*), *dealing with*, *pertaining to*, *on the subject of*, *on*, *about*, or (for a small subset) *over*. The second reason that ABOUT is unusual is that the head noun in ABOUT CNs almost always belongs to a rather restricted set of either abstract nouns or activity nominalizations, while its modifier noun in many cases is also abstract. To illustrate, (4.26) shows a representative sample of CNs derived by ABOUT Deletion.

(4.26)   a. *abortion vote*   *financial report*
            procedural motion   *theoretical claims*
            criminal policy   *linguistic journal*
            tax law   *commercial agreement*
            history conference   *political discussion*

         b. *pastoral art*   *political cartoons*
            disaster flic   *social satire*
            adventure/love story   *academic novel*
            sports magazine

         c.
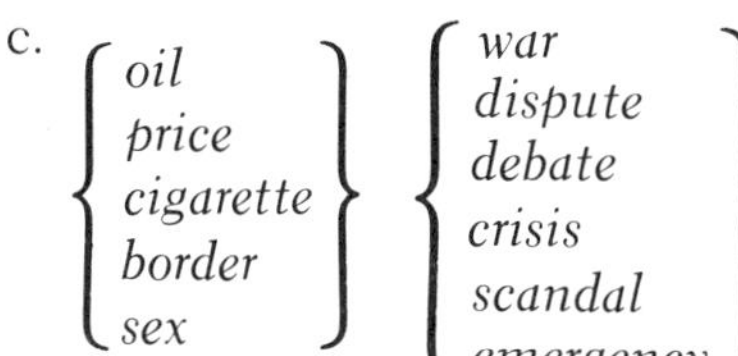

The CNs of (4.26a) are set apart on the grounds that their head nouns all seem to denote either a human speech act of some sort (e.g., *political discussion*), its abstract result (e.g., *tax law* as an abstract mass noun), or a written equivalent of such an act (e.g., *financial report*). This group thus shows a relatively high degree of semantic homogeneity (probably exceeded only by the FROM set just discussed) in comparison with the CN sets generated by other RDPs. The two other major subgroups that fall under ABOUT are the "representational" group in (4.26b) and the "crisis and conflict" group shown in (4.26c); the three divisions indicated here, however, are intended as suggestive rather than systematic ones, since CNs like *love story* in (b) and *border dispute* in (c) might reasonably be viewed as belonging to the "speech act" set in (a).

Although all the CNs in (4.26) may be paraphrased in roughly the same fashion, there are still some differences in preferred periphrastic patterns; thus, as the reader can easily verify, the CNs in (a) can generally be paraphrased along the lines of 'N_2 on the subject of [or on a subject in] N_1', those in (b) by an expression like 'N_2 whose subject is N_1', and those in (c) either by 'N_2 over N_1' (where N_2 denotes an activity like *war* or *dispute*) or by 'N_2 concerning N_1' (where N_2 denotes a state like *emergency* or *crisis*). Whether these differences are primarily due to (and thus predictable from) fundamental semantic distinctions among the head nouns, or are simply a function of lexical idiosyncrasies, is a question that cannot be resolved satisfactorily here; note, however, that in either case, an accurate description of the distinctions would not need to be duplicated in that part of the grammar that derives the CN forms themselves.

A third distinguishing characteristic of these CNs is that although the choice of head noun appropriate to this relationship appears to be relatively restricted on semantic grounds (i.e., to those abstract entities or activities which may be said to have a subject matter or central focus), the choice of modifier is immensely less constrained; in fact, the number of nouns that could denote the subject of a conference, a body of law, a report, or a discussion is virtually unlimited.

Although it remains to be seen just why the predicate ABOUT has certain characteristics that are unusual for an RDP, the productivity of this predicate in forming CNs (due primarily, as we have seen, to the selectional freedom associated with the choice of modifier) constitutes the strongest argument for its inclusion in the RDP set.[16]

---

[16] There is an alternative hypothesis that I would like to bring to the reader's attention, which might eventually explain just why the CNs analyzed here under ABOUT display certain unusual properties. A comparison of these CNs with the four groups of "picture nouns" analyzed in some detail by Warshawsky Harris (1965) reveals a striking degree of overlap in the examples cited, as well as a number of additional similarities between her analysis and

                                    4 CN DERIVATIONS BY PREDICATE DELETION

It is important to recognize at this point that the relationship denoted here by ABOUT has very specific semantic content, and is in no way to be equated to some maximally vague notion of 'x is related to y' or 'x has something (unspecified) to do with y'. The latter notion corresponds to the non-analysis that has appeared with disappointing frequency in studies of compounds in various languages, in which it is claimed that the idiosyncrasies of nominal compounds are so great that **no** semantic regularities can be discerned at all. This treatment of CN forms as **inherently** idiosyncratic (an approach discussed in more detail in Section 3.2) is exemplified by these comments from Jespersen's analysis of English nominal compounds (Jespersen 1942, Volume 6, Chapter 8):

> Compounds express a relation between two objects or notions, but say nothing of the way in which the relation is to be understood. That must be inferred from the context or otherwise. . . . On account of all this it is difficult to find a satisfactory classification of all the logical relations that may be encountered in compounds. In many cases the relation is hard to define accurately [p. 137].
>
> No definite and exhaustive rules seem possible [p. 140].
>
> The number of possible logical relations between the two elements is endless [p. 143].

Unfortunately, an analysis like this one which places no inherent constraints on the formation of CNs amounts to no explanation at all. Two of the most serious drawbacks in such an approach were identified earlier in Section 3.2, where I claimed that such analyses simply cannot account either for the recursiveness of CN formation or for the spontaneous invention and easy interpretation of novel CN forms. Moreover, the paraphrases that this approach suggests as reasonable for CNs can be seen to have no relation at all to the ways in which these forms are actually understood; this was demonstrated by the examples shown in (4.9) (q.v).

---

mine. (These include the notion of an implied subject matter, a preponderance of nominalized verbs and of nouns representing human acts of communication, a wide choice of objects for the underlying verbs, and a number of nouns that could plausibly be analyzed as opaque nominalizations, such as *book*, *idea*, *journal*, and *portrait*.) What this suggests is that at least some of the CNs analyzed here as being derived by ABOUT Deletion might be more appropriately analyzed as CNs derived by nominalization processes instead, from verbs whose objects are preceded in non-CN forms by *about*, *on*, or *over* instead of the more common *of* (cf. *lecture about/discussion of linguistics, law on/regulation of taxation, dispute over/comparison of theories*). Although a derivation of these forms by predicate nominalization rather than by predicate deletion creates its own theoretical complications (which is why it has not replaced the ABOUT analysis in this work), such a derivation would be well worth investigating in a more thorough study of these forms than has been attempted here.

In summary, this "maximally general" and basically agnostic perspective on CNs is in no way equivalent to the very specific analysis of CNs derived by ABOUT Deletion that is proposed here. The former approach is, in fact, more appropriately viewed as a broad **pragmatic** principle than as a syntactic or semantic theory of any substance. Its pragmatic significance follows from the observation cited earlier by Downing (1975) concerning novel CNs: "Because the compounding process is extremely productive, and because compounds are considerably more transparent semantically than novel monomorphemes, compounds are ideally suited to serve as ad hoc names [p. 11]." Let us thus assume, following Downing, that one major function of CNs is to provide names with sufficient semantic transparency that a hearer can readily identify the referent of a novel CN without asking for more information. We must then recognize that for a speaker to construct a CN whose modifier in fact had nothing to do with the head noun would radically interfere with effective communication, since such a choice would lead the hearer away from, rather than towards, correct identification of the intended referent. When we hear an unfamiliar CN, we therefore assume a priori that the modifier has been chosen to represent a relation that **does** obtain between the two surface components (so that the only task remaining is to figure out what that relation actually is). This assumption is, however, essentially a principle of pragmatics that must not be confused with the semantic principles and syntactic rules which make up a **grammatical** theory of CN formation like the one presented in these pages.

4.1.4 Exclusion of LIKE

There are two sets of CN forms whose interpretation regularly appears to include a relationship of similarity or "like-ness," and which therefore suggest that the RDP set ought to be expanded to include a predicate such as LIKE or RESEMBLE. It is my purpose in this section to explain why I have not permitted such an inclusion.

The first set of relevant data consists of CNs which may be used not only in one or more literal senses, but also as either full or partial metaphors. CNs used as full metaphors, such as *coffin nails* (to mean 'cigarettes') or *beehive* (to mean 'place of feverish activity'), were discussed earlier in Chapter 1, where I explicitly excluded the metaphorical readings of such forms from the set of readings that my theory can—and should—predict. On the other hand, CNs used as partial metaphors (i.e., CNs in which only one component is intended metaphorically, such as *mother church*, *handlebar mustache*, and others cited in [4.14]), have been analyzed in this chapter as a subset of CNs derived by BE Deletion; my claim is that such a derivation suffices for associating the surface form with the appropriate range of **literal** readings, and that the metaphorical extension which is possible for forms like these cannot and should not be accounted for by a grammatical analysis, but—more probably—by one or more pragmatic principles concerned with metaphorical extension in general. I shall return to this alternative approach at the end of this section.

The second set of CNs in question requires more extensive discussion. This set consists of a large number of CNs whose surface forms permit a choice of interpretation between their basic reading(s) and an extended reading involving an embedded comparison. For example, the CN *imperial bearing* may be used to mean either 'bearing of an emperor' or 'bearing LIKE the bearing [which is typical] of an emperor'; similarly, *maternal attentions* may have a reading of either 'attentions of a mother' or 'attentions LIKE the attentions [which are typical] of a mother'.[17] Other examples of such double-reading forms are given in (4.27).[18] (Although my study has in general not included CNs whose modifiers are proper nouns or PPAs, such forms are shown here in [c] since they are often characterized by this kind of interpretive flexibility.)

(4.27) a. *imperial bearing* b. *city amenities*
           maternal attentions   *family problems*
           lunar landscape   *summer sunshine*
           professional attitudes *establishment values*
           suburban lifestyle   *preacher talk*
           feminist analyses   *country charm*

        c. *Shakespearean language*
           Chomskyan analyses
           Nixonian politics
           Markovian solutions
           Medean fury

[17] I am grateful to Dwight Bolinger for pointing out to me (personal communication) that to analyze this difference in meaning solely in terms of a predicate such as LIKE is to ignore the notion of typicality that these CNs also regularly carry. Since, however, the use of typical (rather than accidental or infrequent) features as a basis for comparison seems to be required for a wide variety of comparative constructions (e.g., *He's as demanding as my three-year-old* is a comparison based on the **typical** demandingness of the latter), I shall assume that this must follow from some very general pragmatic principle of comparison and so shall not emphasize it further in the present context.

[18] Oddly enough, although both Adj–N and N–N CNs may display this double reading, the derivational types that do so appear quite restricted; specifically, almost all the CNs with this structural ambiguity are derived by HAVE$_2$ deletion or (for the metaphorical types like *hairpin turn* or *queen bee*) by BE deletion. I have no explanation at present for this lopsided distribution.

Because the extended readings of these CNs follow the general pattern of 'N$_2$ which is like N$_1$N$_2$', I shall call these forms, when used on their extended readings, **CN similes.** As is true in regard to CN metaphors, there are two distinct approaches that we must consider in analyzing the double readings that are typical of CN similes, namely, a syntactic–semantic approach and a pragmatically based approach. Each of these theoretical options offers both distinct advantages and somewhat unsatisfactory consequences. Let us consider the syntactic–semantic approach first, since it would appear to be a natural extension of the RDP analysis developed in the present chapter; a pragmatic alternative will be sketched out subsequently.

In order to explain the double readings assignable to each of the CN similes in (4.27) from a standard generative semantic perspective, one could plausibly hypothesize that the CN carrying the extended meaning must have a source structure whose highest predicate is LIKE, whose subject NP is the head noun of the surface CN, and whose object NP is the NP constituent from which the "basic" CN is itself derived. The CN *imperial bearing*, for example, would then have two possible sources: a structure corresponding to *bearing that an emperor has* and another corresponding to *bearing like (the) bearing that an emperor has*. (For present purposes, we may ignore the obvious fact that *imperial bearing* is as many ways ambiguous in principle as any other CN, in order to concentrate instead on just that basic reading which is based on HAVE$_2$.) Note that the semantic structure proposed for the CN with the extended reading includes within it the entire semantic structure of the CN with the basic reading.

Although I will conclude this section by suggesting that a pragmatic explanation for the double readings in question is superior to a syntactic–semantic one, the generative semantic hypothesis just outlined does receive support from a number of syntactic and semantic features of forms like those in (4.27). Since, moreover, these syntactic and semantic features must be accounted for in any complete explanation of CN similes, be it a syntactic or a pragmatic one, it will be worth our while to examine these features in more detail before drawing any conclusions about the best overall explanation for this construction.

To begin with, there are systematic differences in the syntax and semantics of the adjectives that appear in these forms; in the extended readings, the adjective is both predicating and intensifiable, unlike the nominal adjective in the basic reading of the CN. This is illustrated in (4.28)–(4.29), where the (a) readings are the basic ones and the (b) readings the extended ones.

      4 CN DERIVATIONS BY PREDICATE DELETION

(4.28)   a. *In the great Audience Hall, all were moved to respectful
            silence by the force of the*

            $\begin{cases} imperial\ bearing\ [\ =\ emperor's\ bearing\ ] \\ *bearing\ which\ was\ imperial \end{cases}$.

         b. *In the servants' quarters, all were moved to giggles when
            the footman affected*

            $\begin{cases} his\ (extremely)\ imperial\ bearing \\ a\ bearing\ which\ was\ (extremely)\ imperial \end{cases}$.

(4.29)   a. *In that family, maternal attentions [= the mother's
            attentions] to the children are more often punitive than
            supportive.*

         b. *In that family, the older brother's attentions to his siblings
            are* $\begin{cases} more\ maternal\ than\ the\ mother's \\ very\ maternal \end{cases}$.

In addition, the fundamental difference in meaning between the adjectives in the basic and extended readings of CN similes permits us to use the same surface adjective in ways which would normally lead to internal contradiction, as may be seen in (4.30).

(4.30)   a. *The imperial bearing today is more dejected than imperial.*
         b. *The maternal attentions in that family are not maternal at
            all.*
         c. *The lunar landscape in some places is startlingly unlunar.*
         d. *Professional attitudes in some quarters are very
            unprofessional.*
         e. *The fraternal embrace her brother gave her was noticeably
            unfraternal.*
         f. *Feline agility becomes progressively less feline once arthritis
            strikes.*

The grammaticality of these sentences (as compared to the superficially parallel but contradictory *The bright student wasn't bright* or *Those convincing arguments weren't convincing*) is predictable from our hypothesis, since the prenominal adjective in each subject NP in (4.30) is derived solely from an underlying noun (e.g., from *emperor, mother,* or *moon,* for the first three cases), while the homophonous adjective in predicate position must be derived from a markedly more complex source, in which the whole CN forms a compound adjective with *-like* as its final element (e.g., [*imperial bearing*]-*like* or [*lunar landscape*]-*like*).

Since, however, the prenominal and predicate adjectives in each sentence would not be derived from the same semantic structure, they would be predictably nonsynonymous; as a result, the double usage in each line of (4.30) reduces to a partial play on words rather than a real contradiction since the two adjectives are not truly identical but rather only homophonous.

The fact that the adjectives on the extended reading are regularly predicating and intensifiable (unlike the adjectives in normal CNs) is predictable from the proposed hypothesis, in which the semantic structure of the former includes a predicate such as LIKE. That these adjectives are predicating follows from the fact that they are semantically equivalent to well-formed predicate phrases of the form *(is) like x* or *resembles x*; thus, the grammaticality of the (a) sentences in (4.31)–(4.32) follows from the grammaticality of the synonymous (b) sentences.

(4.31)   a. *His bearing was imperial.*
            b. *His bearing was like an emperor's (bearing).*

(4.32)   a. *The desert landscape appeared strikingly lunar.*
            b. *The desert landscape appeared strikingly like the moon's (landscape).*

The intensifiability of the adjectives on the extended reading is also predictable if LIKE is present in semantic structure. Because the relation of "LIKE-ness" is an intensifiable one, lexicalizations of LIKE regularly appear with degree adverbials; thus, we can say that x is *very like* y, *quite like* y or *slightly like* y, regardless of what x and y are. As a consequence, the fact that degree adverbials are grammatical in such expressions as *extremely imperial bearing, a somewhat lunar landscape,* or *very maternal behavior* may be explained by claiming that they are modifying the intensifiable predicate LIKE in semantic structure, rather than the nonintensifiable CNs whose modifier nouns form the morphological stems of these adjectives. (Note that such degree adverbials are inadmissible with the basic readings of these CNs; for example, in speaking of the Judaean desert, one could describe its *very lunar landscape,* but in speaking of the moon, one would have to omit the *very.*)

When we consider CN similes with an N–N structure rather than an Adj–N structure, we find once again that two readings are possible and that these readings display systematic differences. The fact that an extended reading with LIKE is possible for N–N forms as well is shown in (4.33), where the contexts have been contrived to force the extended reading instead of the basic one regularly associated with each CN. (Note that quotation marks are often used as a graphic marker to signal that the intended reading is the extended one, as in the examples of [b] and [d].)

(4.33)   a. *For a farm community, they certainly have many **city amenities**.*
          b. *The chairperson was reluctant to talk about the "**family problems**" that were troubling the department.*
          c. *That Roman poet certainly shows a **feminist perspective** in many of his verses.*
          d. *The burst of "**summer**" sunshine was particularly welcome on that cold day in March.*
          e. *She was infuriated by her working class uncle's **establishment values**.*

The semantic difference between regular CNs and those isomorphic forms with a LIKE reading that are shown in (4.33) is demonstrated by the fact that N–N forms on the extended reading, but not on the basic one, may usually be paraphrased by an N_1-*type* N_2 construction. For example, we could easily substitute the phrases *city-type amenities, family-type problems*, or *a feminist-type perspective* in the first three sentences of (4.33), but not in contexts where the basic reading is intended. (Thus, consider how odd it would sound if parents were to consult a social worker for their *family-type problems* rather than for their *family problems*, or if we would seek out the *summer-type sunshine* rather than the *summer sunshine* on a bright August day.)

The facts just discussed seem to support the hypothesis that the semantic ambiguity of the CNs in (4.27) and (4.33) can be analyzed in terms of a **structural** ambiguity, that is, in terms of two related source structures whose primary difference is the presence or absence of LIKE as a highest predicate. If we observe further that (*a*) CNs with this LIKE reading are not at all uncommon; and (*b*) the LIKE reading of these forms is easily understood despite the absence of any surface morpheme that expresses this LIKE relationship, we might be tempted to propose an additional hypothesis to the effect that the predicate LIKE must also be a member of the RDP set, that is, that LIKE is present in the underlying structure of many CNs but is regularly and recoverably deletable in the course of their derivations. There are, however, several compelling reasons to reject any hypothesis that would extend the RDP set to include a predicate LIKE.

To begin with, we must note that there is something peculiar about the fact that the extended readings of forms like those cited earlier are regularly parasitic on the basic reading(s) of the normal CN. That is, if we would peel off that part of each relevant paraphrase that contains the predicate LIKE, we would still be left with a CN which itself must be accounted for; there is no way that we can fully understand the meaning of a CN simile on a LIKE reading without a correct analysis of the CN contained within it. For example, to use the CN *summer sunshine* in its extended sense of 'sunshine like summer sunshine' we must already know that the basic reading of the form is, say, 'sunshine in summer' and not 'sunshine causing summer' or even 'sunshine remembered in summer'; similarly, to account for the extended sense of *imperial bearing*, we must know that the basic form itself is used to mean 'bearing that an emperor has' rather than 'bearing that an emperor envies' or 'bearing appropriate in the presence of an emperor'.

This parasitic or piggyback quality is one of the most important reasons for excluding a predicate LIKE from the RDP set. LIKE cannot be analyzed as an RDP since its deletion would produce a reading which itself includes a reading based on the deletion of another RDP; this is quite different from the analysis of true RDPs, which generate CNs by derivations that are mutually exclusive and independent of each other. This is not to say that a given form might not be analyzable under more than one RDP since this is in fact not unusual, but rather that no set of CNs derived by the deletion of one RDP regularly and **invariably** includes a reading based on the deletion of any other RDP. (Note that the lengthier forms which may be derived by the cyclic application of various· RDP Deletion rules are not relevant to this discussion, since it is only those two-element CNs comprising a head and just one prenominal modifier which can be legitimately compared to the forms like *imperial bearing* and *city amenities* which are under scrutiny here.)

It may have occurred to the reader that this "piggyback" reading could perhaps still be accounted for on the basis of the independently motivated treatment of CN formation as a cyclic process. Thus, if LIKE **were** included among the RDPs, the first cyclic application of the relevant rules would produce the innermost CN (i.e., the CN on its basic reading), while the derived LIKE reading would be formed when the rules (including a rule of LIKE Deletion) would apply on the next higher cycle. The derivation that this approach entails is sketched in linear form here:

(4.34) a. **First Cycle:** Compound Adj Formation, HAVE Deletion
bearing which is LIKE *bearing which emperor* HAS
bearing which is LIKE *bearing which is emperor-had*
bearing which is LIKE *emperor-had bearing*
bearing which is LIKE *emperor* Ø *bearing*
bearing which is LIKE *imperial   bearing*

b. **Second Cycle:** Compound Adj Formation, LIKE Deletion
bearing which is LIKE *[imperial bearing]*
bearing which is [imperial bearing]-LIKE

$$[\text{imperial bearing}]\text{-}\textsc{like} \text{ bearing}$$
$$[\text{imperial bearing}] \quad \emptyset \quad \text{bearing}$$
$$\text{imperial} \quad \emptyset \quad\quad\quad \text{bearing}$$

At least two objections to this analysis must be raised, however. The first objection is that it would require an ad hoc rule, not needed anywhere else in CN formation, to delete the leftmost occurrence of the head noun and thus avoid the redundancy of the duplicated structure produced by LIKE Deletion (i.e., to change the *imperial bearing bearing* of the penultimate line of [4.34b] into the acceptable *imperial bearing*). The second, and stronger, objection is that even if an RDP LIKE could be deleted on the **second** application of the relevant rules, as sketched in (4.34b), this analysis could not explain the fact that the alleged RDP LIKE cannot be deleted on the **first** cyclic application of these rules, as the data in (4.35)–(4.37) demonstrate. If LIKE were an RDP, it should both form compound adjectives and undergo deletion from these adjectives when they are in prenominal position, leaving a noun behind as the first element in a CN. That LIKE does not undergo such deletions without change in meaning, however, is shown by these examples:

(4.35)   a. *professors who are like children*
              b. *child-like professors*
              c. *child professors*

(4.36)   a. *trees which are like pines*
              b. *pine-like trees*
              c. *pine trees*

(4.37)   a. *arguments which are like wars*
              b. *war-like arguments*
              c. *war arguments*

If LIKE shared the syntactic properties of other RDPs, its deletion from the compound adjectives in the (b) examples would produce CNs which are synonymous (on one of their regular readings) with both the (b) and (a) structures that precede them; as the reader can see, however, the CNs in (c) do not have the readings of either the (a) or (b) expressions from which they were presumably derived. We may therefore conclude that even the cyclic application of CN formation rules cannot be used to justify the inclusion of LIKE among the RDPs.

If, however, we exclude LIKE from the RDP set, we must still explain how it is that the extended readings of so many CNs **seem** to include a predicate such as LIKE in addition to their basic readings, despite the fact that there is no overt marking to signal the difference. That is, if we still wish to hypothesize that the semantic structure of forms like those in (4.27) and (4.33) includes a predicate such as LIKE (even if it is not an RDP), what process or principle can account for the fact that its disappearance in the course of a derivation evidently does not preclude its being semantically recoverable? And if we do not make such a hypothesis, how else are we to account for the very real differences in interpretation that we have seen are possible? (The reader should note that the problem is not a function of expressing semantic relationships in terms of underlying predicates; in a theory which chose instead a system of semantic interpretation rules for associating meanings with CN and other forms, the unanticipated fact that such rules would have to assign these double readings systematically would still have to be explained.)

I believe that the correct explanation for these double readings must be sought in the broader context of "extended readings", that is, in the same context in which we place "metaphorical" language use in general. It appears to be true that virtually all linguistic forms can slide in actual discourse from requiring the most concrete or literal interpretation possible to permitting the most metaphorical, without any **overt** marking of just how extended or metaphorical the reading is intended to be. This flexibility is therefore characteristic not just of CNs, but rather of a vast array of syntactic constructions in any language; to take but a few examples from English, consider the boldfaced forms in (4.38):

(4.38) a. *The impatient clerk absolutely **growled** at me.*
    b. *You have brought **sunshine** into my life with your good news, Roderick!*
    c. *They were reduced to living in **a hole in the wall** when they both lost their jobs.*
    d. *Her expression **blared out** smugness to all but the most obtuse.*
    e. *"Be more subtle," advised the editor. "You don't have to make your points **with a sledge-hammer,** you know."*

The fact that we easily interpret these expressions in a nonliteral manner despite the absence of overt "instructions" to do so is an indication that readings of particular lexical items or syntactic forms which extend beyond their literal (or customary) referents are so common as to make a **structural** analysis of such extensions (such as the analysis just discussed, based on a LIKE predicate) hopelessly inadequate to the task. Rather it seems likely that a small set of pragmatic principles will provide the best explanation for the fact that we regularly use and interpret a great variety of linguistic forms as instances of semantic extension; whether we classify this extension as a metaphor, a simile, or as any other rhetorical device (e.g., metonymy, synecdoche, or hyperbole), it remains the case that the extension is not signaled by any overt linguistic form but is instead inferred, with greater or lesser accuracy, from contextual clues. Moreover, because the appropriate interpretation is governed in these cases by the immensely complex and variable situation which we refer to (often rather blithely) as "the context," these extensions of reference cannot be formalized in the same manner (if indeed they can be formalized at all) as those other, invariant and predictable aspects of meaning which we seek to capture in specific derivational analyses.

Returning now to the particular problem of CN similes like those of (4.27) and (4.33), I propose that the pragmatic principle which permits us to assign appropriate extended readings in these cases might be stated roughly as follows: "If no literal meaning of a CN makes sense (or is relevant, or applicable, or internally consistent) **in a given context,** interpret a CN of the form N_1N_2 on a reading comparable to 'N_2 which is like N_1N_2'."

The principle just proposed is intended to reflect two crucial characteristics of CN similes. The first is that the extended reading does not assign to the prenominal modifier a reading comparable to 'N_i-*like*' (where N_i is the prenominal modifier of the whole CN) but rather a more complex—and inherently ambiguous—reading of 'CN_i-*like*'." For example, the extended reading of *city amenities* illustrated earlier in (4.33a) does **not** carry the meaning of '*city-like amenities*' in the strict sense of 'amenities which are like cities' but rather the more complex meaning of 'amenities which are like city amenities', in which the real comparison is to a full CN rather than a single, underived N. There is thus a special recursive quality to CN similes which is **not** present in normal similes but which must be captured directly and explicitly in the way we word the relevant pragmatic principle.

This example leads us naturally to the second essential characteristic of CN similes which our pragmatic principle must reflect, namely, that the extended interpretation of these forms is precisely as flexible (i.e., as ambiguous and/or vague) as is the CN form on its nonextended reading(s). Thus, when we use the form *city amenities* as a CN simile to indicate that a certain community has what we could also call *amenities which are like city amenities,* we leave the interpretation of the embedded CN *city amenities* just as ambiguous and vague as any other nonextended CN, so that the hearer must go through the same process of disambiguation and specification of referent as is required to understand all occurrences of nonlexicalized or unfamiliar CNs in natural discourse. This fact follows, however, from the recursive nature of CN similes in general, a property which is reflected structurally in the pragmatic principle proposed here by the inclusion of a full—but unanalyzed—CN within the extended interpretation.

Other pragmatic principles must still be formulated to account for the sets of CNs (discussed at the beginning of this section) which are used as either full or partial metaphors, a type of semantic extension which is related to, but yet quite distinct from, the CN similes which have been the primary focus of this section. I have no intention of attempting to spell out such a principle (or principles), not least because the task is one which has defeated—or at least confounded—far better minds than mine, but also because its accomplishment is not at all crucial to the conclusion of this section. Instead I will simply reiterate my claim that CNs which are used as full or partial metaphors should be derived by regular CN formation rules, even though such a derivation does not reflect directly the possibility of metaphorical extension; instead, a pragmatic principle of quite general application must account for the perennial possibility of assigning a metaphorically extended interpretation to a particular CN form.

One way in which this principle would simplify grammatical description is by permitting a single derivation (or, more accurately, a single set of derivations) to suffice for every CN form which may be used with **both** literal and metaphorical senses (e.g., *horse race, snake pit, coffin nails, bottleneck, paper tiger, bamboo curtain, tidal wave*); although some of these forms have lexicalized senses when used nonliterally, others may shift back and forth in a most productive fashion without requiring our having to know any lexicalized information at all. (For example, if a newspaper reported that *a tidal wave of opposition to the "air pollution" in the Senate Chamber dealt a mortal blow to the pending tax bill*, it would be using the CNs *tidal wave, air pollution*, and *mortal blow* on metaphorical but basically nonlexicalized readings.)

We may note also that the same pragmatic principle could account for the set of CNs in which only the head noun is intended metaphorically; ignoring for present purposes the fact that the following forms are actually lexicalized CNs, we may illustrate this pattern with CNs such as *cannon fodder, road hog, jailbird, sea horse*, and *brainstorm*. Similarly, CNs in which only the modifier is intended metaphorically could be derived by regular CN formation processes, with the semantic extension involved accounted for not by the syntactic derivation per se but rather by the pragmatic principle(s) concerned with metaphors in general; examples of this pattern include *hairpin turn, queen bee, handlebar mustache*, and *parent organization*. (My assumption that syntactic derivations cannot and should not be made to account for metaphorical extensions of either entire CNs or their component parts is reflected in the fact that

CNs like these last four have been analyzed in this chapter simply under BE, with no distinction made in their derivational histories to separate them from other CNs that are derived by BE Deletion but which lack such metaphorical interpretations.)

If we require of our grammar that it provide only a single derivation for each literal reading of a CN, and then permit general pragmatic principles to account for whatever rhetorical liberties individual speakers may take with any such reading (including, but not limited to, the use of CN similes and metaphorical extension), we shall avoid an immense amount of needless duplication in our description of the semantic properties that may be associated with particular CN forms. Moreover, the fact that semantic extensions are subject to tremendous individual and contextual variation, in ways which appear to be potentially unlimited and hence unpredictable, provides additional motivation for an approach which makes no attempt to specify every possible extended reading of a given CN form. Finally, an appeal to independently motivated pragmatic principles to account for extended interpretations of CNs would reflect the fact that these semantic extensions are not peculiar to the grammar of CNs alone but rather are related to more fundamental principles of language use whose domain of application extends far beyond the area of CNs with which we are concerned here.[19]

My primary purpose in this section has been to explain why no predicate such as LIKE or RESEMBLE has been included within the set of RDPs, despite the fact that a relationship of "like-ness" does seem to be a regular part of the interpretation of both CN similes and CN metaphors. I hope to have shown that the arguments against such an inclusion are convincing ones, although I have done little more than suggest the direction in which we might seek a more satisfactory explanation. It is, of course, far simpler to invoke pragmatic principles in an attempt to solve syntactic dilemmas than it is to integrate such principles explicitly and successfully within a complete description of CN forms. Moreover, although a pragmatic approach appears to avoid some of the pitfalls of the syntactic–semantic approach explored in the first part of this section, there remain some puzzling questions whose resolution may or may not

---

[19] Although my own conclusions with respect to the "LIKE" readings of CNs were reached independently, my discussion here of the larger context of metaphorical usage owes much to the insights of Reddy (1969). Readers interested in the problems posed by questions of metaphor and reference (especially "customary" versus extended) are referred to this article for an excellent and provocative discussion of these issues. Reddy argues persuasively that metaphor is a far more pervasive phenomenon in natural language than many linguists have recognized, and that an analysis in terms of unusually extended **reference** of a term is superior to earlier theories based on violations of predetermined selectional restrictions.

be possible within a purely pragmatic account. Just one of these questions is how a pragmatic explanation of CN similes could account for the specifically syntactic and semantic regularities that are documented in the data shown earlier in this section. I know of no answer to this question at present.

4.2 Derivations

4.2.1 Derivational Progressions

The derivations required for the formation of CNs by predicate deletion are presented in chart form in Table 4.1 (pp. 119–120) and in the form of labeled trees in Table 4.2 (pp. 121–128). As indicated earlier, these derivations are divided into three types according to the morphological shape of the compound adjective formed in an intermediate step; these three types correspond to:

1. CNs whose underlying predicate undergoes passivization and thus is incorporated into the compound adjective in a passive participle form (e.g., *stress caused by heat* > *heat-caused stress* > *heat stress* > *thermal stress*);
2. CNs whose underlying predicate does not passivize and thus is incorporated as an active participle (e.g., *clock which makes music* > *music-making clock* > *music clock* > *musical clock*);
3. CNs whose underlying predicate surfaces in English not as a verb but as a preposition (e.g., *note in margin* > *in-margin note* > *margin note* > *marginal note*).

There is a corresponding difference in syntactic origin of the prenominal modifier in each group; in the first group it is derived from the subject NP of the relative clause, in the second group from the direct object NP of the relative clause, and in the third group from the object of the preposition in the relative clause.

The derivations in Tables 4.1 and 4.2 include three transformational processes that are essential to the formation of CNs: Compound Adjective Formation, Recoverable Predicate (RDP) Deletion, and Morphological Adjectivalization. Since each one will receive individual attention in the sections to follow, this section will be devoted to a number of conditions and qualifications pertaining to other aspects of the derivations.

To begin with, the CNs used as examples in each of the derivations are of an Adj–N form rather than an N–N form so that as many of the relevant rules as possible may be illustrated; the derivation of comparable N–N forms would be identical to those shown here, with the exception

    4 CN DERIVATIONS BY PREDICATE DELETION

TABLE 4.1
Derivations by Predicate Deletion

Type (a): N₂ ## N₁ V N₂ ##

First cycle	*stress ## **CAUSE** heat stress ##*
1. Lexical Insertion	*stress ## cause heat stress ##*
2. Case Prep Spelling	*stress ## cause by-heat of-stress ##*
3. Passive	*stress ## be caused of-stress by-heat ##*
4. Compound Adj Formation	*stress ## be of-stress heat-caused ##*
5. Copula Insertion	—
Second cycle	
6. Relative Pronoun Formation, WH- Fronting	*stress ## which ## be heat-caused ##*
7. WH-*be* Deletion	*stress heat-caused*
8. Predicate Preposing	*heat-caused stress*
9. RDP Deletion	*heat stress*
10. Morphological Adjectivalization	*thermal stress*

Through Step 8:   *battery-operated toys*
                             communist-backed insurgents
                             church-advocated spending

Through Step 9:   *diaper rash* (**CAUSE**)
                             apple core (**HAVE**)
                             daisy chains (**MAKE**)

Through Step 10: *viral infection* (**CAUSE**)
                             feminine intuition (**HAVE**)
                             molecular chains (**MAKE**)

Type (b): N₂ ## N₂ V N₁ ##

First cycle:	*clock ## **MAKE** clock music ##*
1. Lexical Insertion	*clock ## make clock music ##*
2. Case Prep Spelling	*clock ## make by-clock of-music ##*
3. Passive	—
4. Compound Adj Formation	*clock ## music-making by-clock ##*
5. Copula Insertion	*clock ## be by-clock music-making ##*
Second cycle:	
6. Relative Pronoun Formation, WH- Fronting	*clock ## which ## be music-making ## ##*
7. WH-*be* Deletion	*clock music-making*
8. Predicate Preposing	*music-making clock*
9. RDP Deletion	*music clock*
10. Morphological Adjectivalization	*musical clock*

Through Step 8: *fish-eating dinosaurs*
                             woman-hating editors
                             meaning-bearing elements

Through Step 9: *disease germ* (**CAUSE**)
                             windmill (**USE**)
                             apple cake (**HAVE**)

Through Step 10: *professorial friends* (**BE**)
                             microscopic analysis (**USE**)
                             pictorial atlas (**HAVE**)

TABLE 4.1 (*Continued*)
Derivations by Predicate Deletion

Type (c): N_2 ## N_2 Prep N_1 ##	
First cycle	*note* ## **IN** *note margin* ##
1. Lexical Insertion	*note* ## *in note margin* ##
2. Case Prep Spelling	—
3. Passive	—
4. Prep Phrase Formation,	*note* ## *be note in margin* ##
Copula Insertion	
5. Compound Adj Formation	*note* ## *be note in-margin* ##
Second cycle	
6. Relative Pronoun Formation,	*note* ## *which* ## *be in-margin* ## ##
WH-Fronting	
7. WH-*be* Deletion	*note in-margin*
8. Predicate Preposing	*in-margin note*
9. RDP Deletion	*margin note*
10. Morphological Adjectivalization	*marginal note*

Through Step 8: *between-meal snacks*
                    offshore wells
                    underage customers

Through Step 9: *desert rat* (**IN**)
                    guest house (**FOR(FOR)**)
                    tax law (**ABOUT**)

Through Step 10: *polar bear* (**IN**)
                    avian sanctuary (**FOR**)
                    linguistic journal (**ABOUT**)

that the last (optional) rule of Morphological Adjectivalization would not apply.

Another point to note is that the formulation chosen for the Passive Transformation in the Type (a) derivation is in most respects a traditional one. Although it is quite possible that subsequent research will substantiate an alternative formulation of this process, I do not expect that the adequacy of the derivations proposed here will be materially affected. This is because the crucial rules in these derivations apply only after the elements in the relative clause appear in the passive configuration (i.e., with subject and object NPs reordered, and with appropriate modifications of the verb), and I know of no evidence to suggest that the means by which that configuration is reached would be relevant to the operation of subsequent rules.[20]

---

[20] For arguments that passivization may not be at all involved in the derivation of compounds, see Newmeyer 1975, in which it is proposed that compound formation, like predicate raising and nominalization, is a precyclic process. For criticism of this proposal, see footnote 28 in this chapter and Section 5.3 in the next.

I. Type (a): Passive Participle
Example: *stress caused by heat* > *heat-caused stress* > *thermal stress*

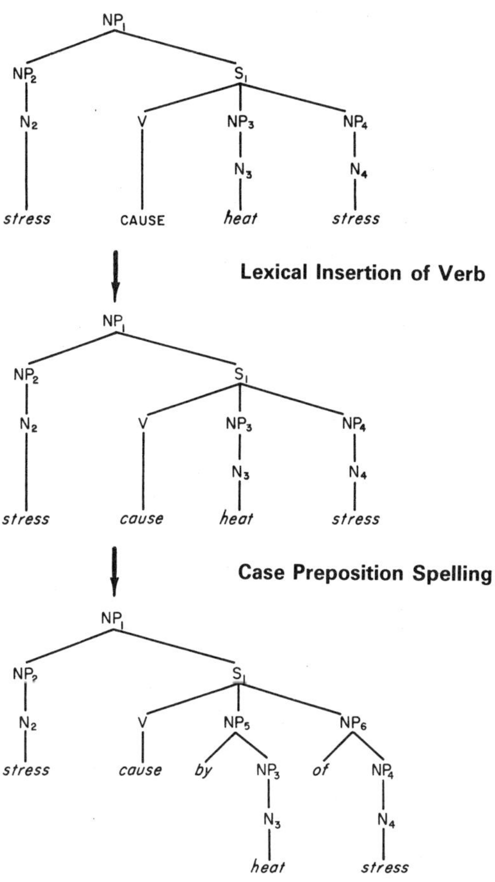

(Continued)

Passive
Compound Adjective Formation
Relative Pronoun Formation and
WH- Fronting

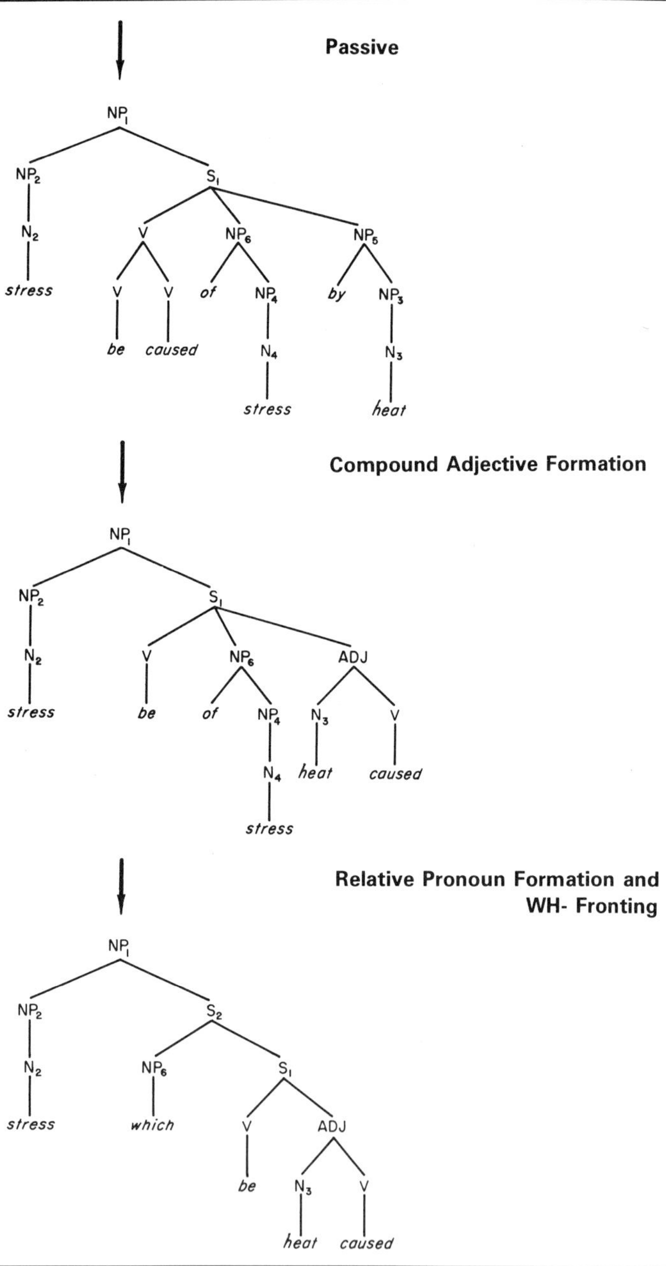

(Continued)

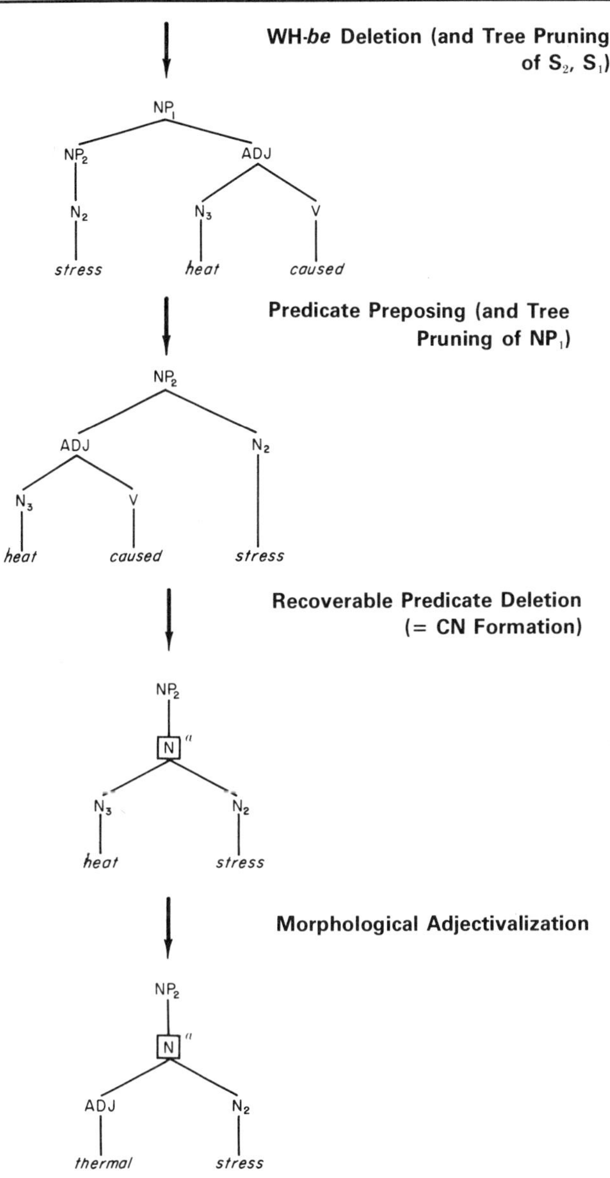

" A box surrounding an N node (as in the trees on these and later pages) is intended as a graphic reminder that that N dominates a CN construction; the box has, however, no theoretical import.

(Continued)

II. Type (b):   Active Participle[b]

  Example: *clock making music* > *music-making clock* > *musical clock*

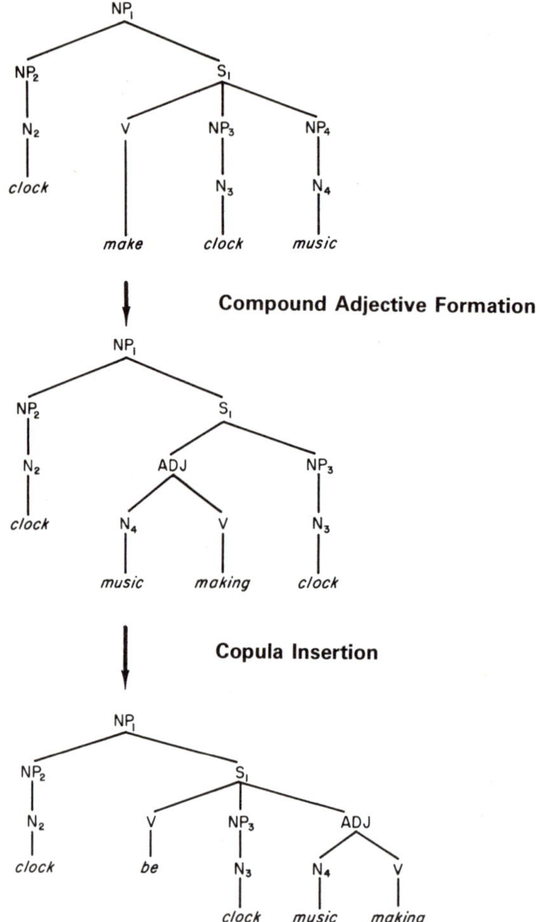

[b] The first two steps in this derivation (Lexical Insertion of Verb and Case Preposition Spelling) have been omitted here since the case prepositions, governed by the output of the first rule and inserted by the second, play no role in the rules of this derivation. To enhance visual clarity, therefore, these markers are not represented in the following trees.

(Continued)

Relative Pronoun Formation and
WH- Fronting
NP₁
NP₂
N₂
clock
S₂
NP₃
which
S₁
V
be
ADJ
N₄
music
V
making
WH-be Deletion (and Tree Pruning
of S₂, S₁)
NP₁
NP₂
N₂
clock
ADJ
N₄
music
V
making
Predicate Preposing (and Tree
Pruning of NP₁)
NP₂
ADJ
N₄
music
V
making
N₂
clock
Recoverable Predicate Deletion
(= CN Formation)
NP₂
N
N₄
music
N₂
clock

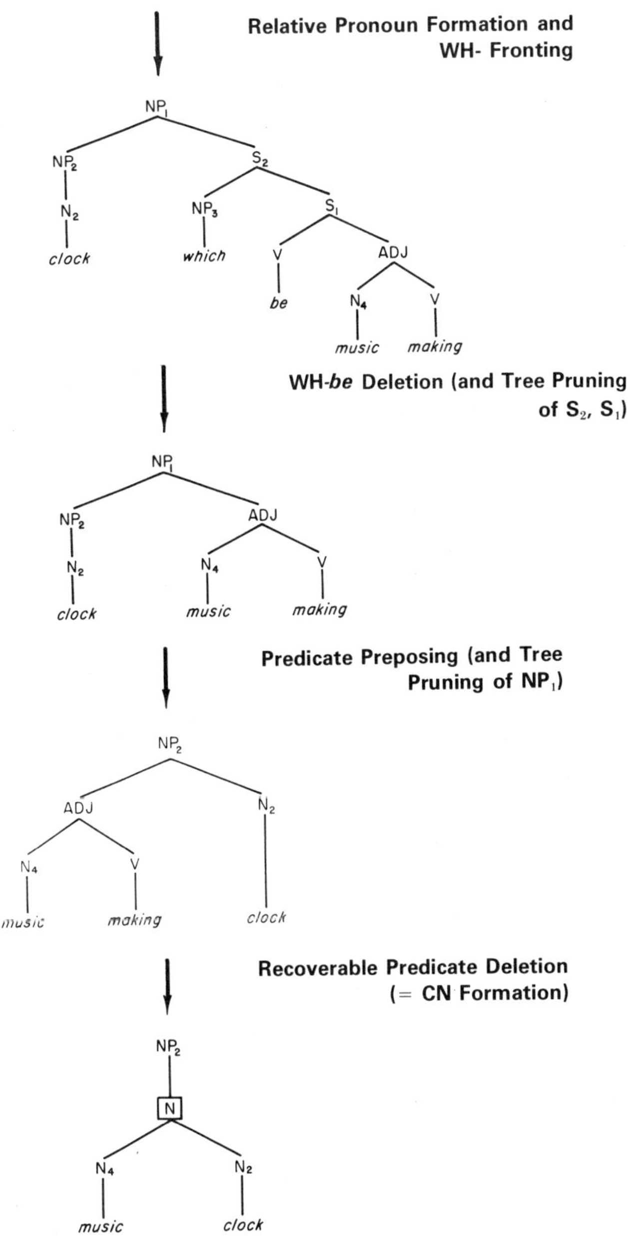

(Continued)

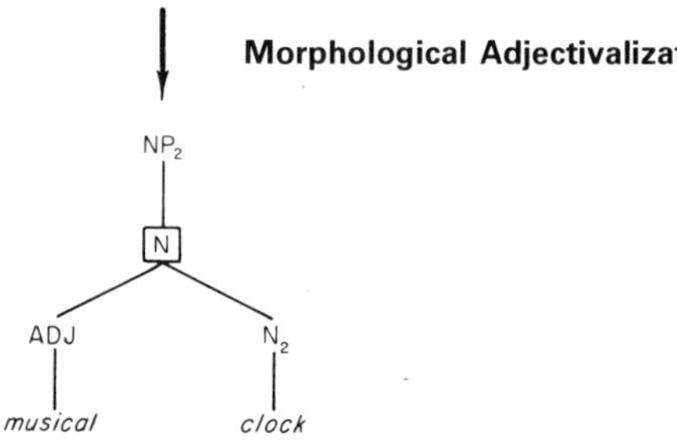

III. Type (c):   Prepositional
    Example: *note in margin > in-margin note > marginal note*

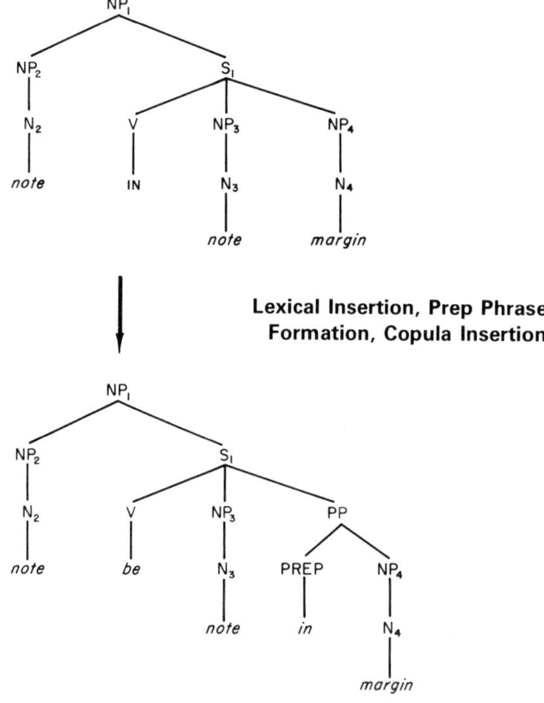

(Continued)

Compound Adjective Formation
Relative Pronoun Formation and
WH- Fronting
WH-*be* Deletion (and Tree Pruning
of S₂, S₁)

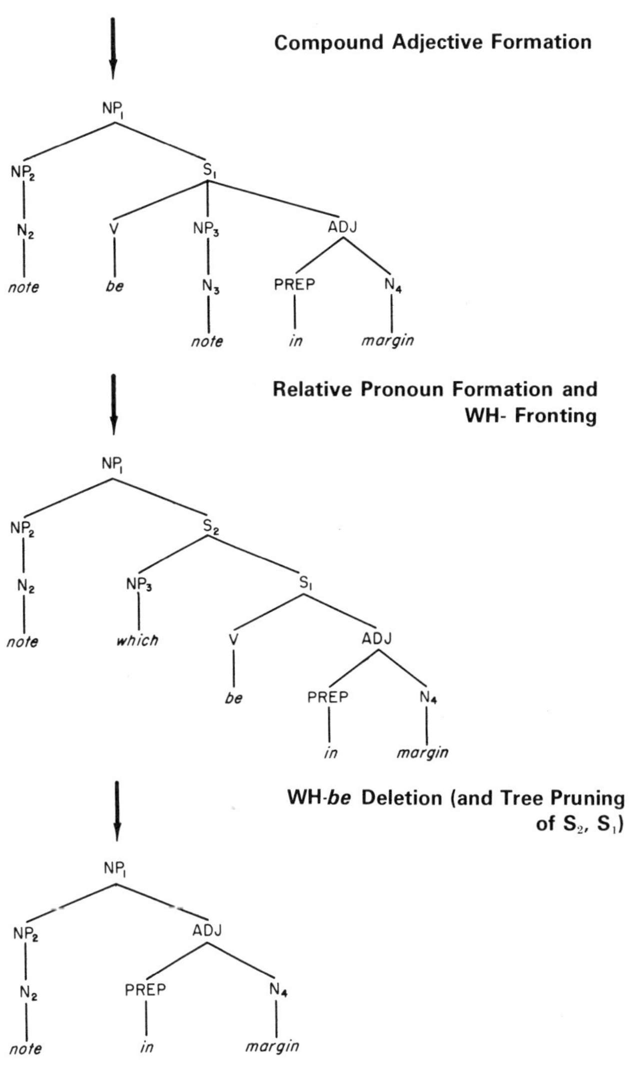

(Continued)

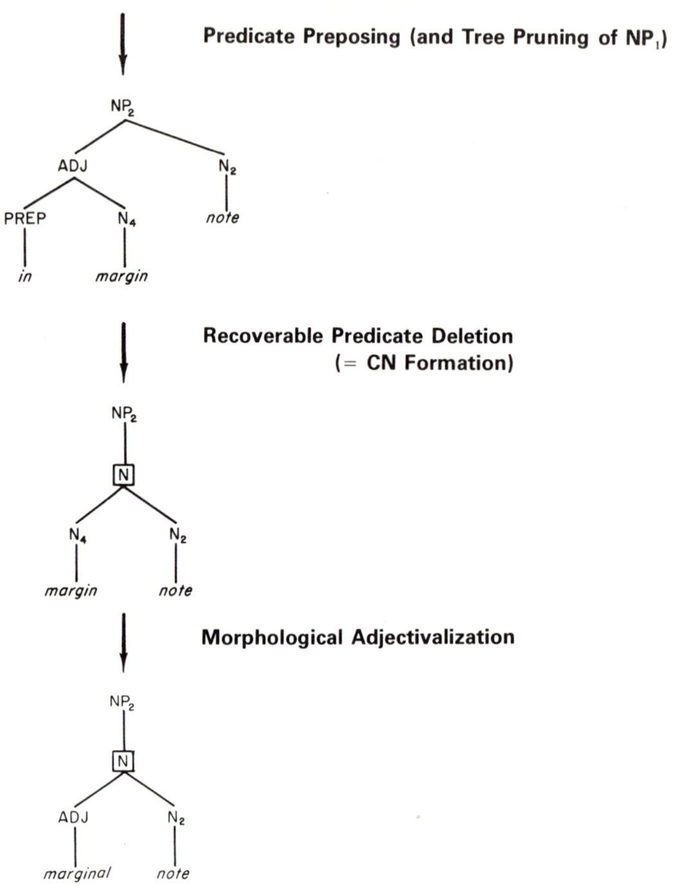

Although a special rule of Copula Insertion has been included in the Type (b) derivation following Compound Adjective Formation, the evidence is less than conclusive that such an inclusion is necessary here. (Note that, in contrast, the relative clauses of Type (a) get their copula via Passive, and those of Type (c) via an independently motivated rule related to predicates lexicalized in prepositional form.) Its dubiousness stems from the fact that the Type (b) compound adjectives that seem to demand a copula never actually surface in predicate position; thus, the relative clauses *a clock which is music-making* or *editors who are woman-hating* are ungrammatical, although semantically not anomalous. It appears to be the case that these adjectives may not remain in predicate position of

the relative clause, but must instead be preposed to a position in front of
the head noun by the obligatory (in these cases) transformation of Predicate Preposing which follows Relative Clause Reduction.[21]

Note, however, that the latter rule also goes under the name of WH-*be*
Deletion, implying that both a WH- form **and** a copula must be present
for the structural description to be met. But the evidence from these
compound adjectives, at least, suggests that this rule might be better
known as something like WH(*be*) Deletion, to cover those cases in which
there is no compelling reason to insert a copula. McCawley has observed
(personal communication) that this conclusion is supported by a second
group of data, namely, reduced relative clauses with stative verbs, such as
people resembling Nixon or *investors owning blue-chip stocks*. Unless the
rule of Relative Clause Reduction is reformulated to operate on structures without a copula, these phrases would have to be derived from the
same antecedent structures that would underlie the ungrammatical
**people who are resembling Nixon* and **investors who are owning blue-
chip stocks*, that is, from structures with an unmotivated (and, indeed,
prohibited) copula which could never reach the surface.

It is not my purpose here to propose and motivate a reformulation of
Relative Clause Reduction, but simply to point out that the Copula
Insertion rule shown here may turn out to be a dispensable step for
the Type (b) derivation in particular. Although it is clear that a full
account of these derivations must somehow relate the forms of the Type
(b) verbs in unreduced relative clauses to their corresponding participial
forms in reduced ones, the manner in which this is to be done is a
problem that need not be solved here.

The constituent structure indicated in each stage of the derivations in
Table 4.2 has been drawn to provide as much relevant detail as possible.
In addition, subscripts on node labels have been included to help the
reader follow changes in constituent structure (although the specific
numerals are without theoretical significance). Determiner nodes that
might emerge as sister nodes to CNs in surface structure have not,
however, been included in these trees. Such detail has been omitted since
virtually no systematic research has been done to date on the sources of

---

[21] An apparent exception to this rule (drawn to my attention by Georgia Green) is a form
like *something music-making* which we would expect to be ungrammatical; it is, however,
irrelevant to the present discussion since the postnominal position of the adjective is a
function of the pronominal head *something* and not of the compound adjective itself. This
is shown by the fact that all NPs with indefinite pronominal heads like *something, someone,*
or *someplace* must have modifying adjectives appear after, rather than before, that head (cf.
some funny thing but *something funny, some clever person* but *someone clever, some new
place* but *someplace new*).

determiners in NPs whose heads are CNs; in particular, analyses must yet be carried out to determine the interaction of determiners originating in the highest NP (e.g., *the/some/three/all/any/those notes which are in margins*) with those originating in any of the NPs contained in the underlying relative clause (e.g., *notes which are in the/some/three/all/any/those margins*).[22] As a consequence, the addition of DET nodes to the trees of Table 4.2 would provide no new information to the reader, especially since the rules as formulated have no effect on such nodes. Nevertheless, it must be recognized that the present analysis can only be viewed as incomplete as long as determiner-related phenomena remain unexplored.

The reader should take note of the fact that my ordering of Lexical Insertion as the first rule in the cycle in Tables 4.1 and 4.2 is based on a number of assumptions that are not shared by all transformationalists, or even by all generative semanticists; these include the following assumptions:

1. At least some rules are ordered in a transformational derivation.
2. There is a rule of Case Preposition Spelling.
3. The case prepositions supplied by this rule are lexically determined.
4. A rule of Passive applies in the derivations of Type (a).
5. The rules of Lexical Insertion, Case Preposition Spelling, and Passive have the intrinsic ordering just cited, on the grounds that the first feeds the second, and the second the third.

Since my intention here is simply to acknowledge these assumptions rather than to argue for them, this brief discussion of ordering may be concluded with a reminder to the reader of two significant points brought out earlier in footnote 3, namely, (*a*) generative semantic theory does not yet permit us to specify with any great precision or confidence just where in the derivation of these forms lexical insertion of predicates **must** take place; and (*b*) an ordering of Lexical Insertion later in the cycle would in fact make these derivations work more rather than less smoothly. Both these reasons, together with the assumptions already specified, lie behind the ordering shown in Tables 4.1 and 4.2.

As a last general comment concerning these derivations, it is important to remember that although Tables 4.1 and 4.2 show the derivation of

---

[22] The only study that I know of which directly addresses the problem of determiners in complex NPs of any sort is the relatively brief article by Coursaget-Colmerauer (1973), which raises a number of pertinent questions concerning constraints on the distribution of determiners in French nominalizations (e.g., *une/toute/cette/les/quelques/plusieurs intervention(s) de l'O.N.U.*, *l'/*une/*cette arrivée du printemps*) and thus has at least some relevance to the analysis of CNs based on nominalizations, like those to follow in Chapter 5.

          4 CN DERIVATIONS BY PREDICATE DELETION

the simplest forms of CNs (i.e., two-element CNs), the demonstrably recursive nature of CN formation entails that the rules of these derivations could apply repeatedly with the input at each new stage including a CN produced by the previous stage. We must therefore regard all the crucial rules (especially Compound Adjective Formation) as cyclic ones. (A fuller discussion of the issue of cyclicity is provided in Section 5.3.)

Let us now turn to the detailed analysis of each of the three crucial processes which together produce the CNs of this chapter: Compound Adjective Formation, Recoverable Predicate Deletion, and Morphological Adjectivalization.

4.2.2 Compound Adjective Formation

Of the many types of compound adjectives in English, we shall be concerned here only with those that combine a predicate element (participle or preposition) with its nominal object, since only those types may underlie CNs.[23]

There is no single transformation that may be written to generate all the types of [predicate + noun] adjectives that occur in English, since the form of the predicate element varies from type to type, as does the form of the compound adjective that is produced. The differences in input and output may be seen in (4.39); the strings on the left represent typical inputs of each type, while the strings on the right show typical compound adjectives that are formed for each type.

(4.39) a. *wave be caused by tide* > *wave be tide-caused*
     b. *clock makes music*   > *clock (be) music-making*
     c. *note (be) in margin*   > *note (be) in-margin*
     d. *town (be) free of pollen* > *town (be) pollen-free*

Of these four types, only the first three will be analyzed in this work since they alone may be sources of the prenominal modifier of a CN; for an as yet undetermined reason, the predicates that occur in the *pollen-free* type

---

[23] For a fuller classification of the various occurring types of English compound adjectives, see Jespersen 1942:Volume 6, Sections 9.5–9.6; Marchand 1969:84–95; and especially Meys 1975. Numerically prefixed adjectives which might be analyzed as including a noun and a predicate as well as a numeral (e.g., *monolingual, polychromatic, trisyllabic*) will be excluded from discussion here despite their superficial similarity to nominal adjectives (cf. *lingual, chromatic, syllabic*) since their semantic composition and syntactic behavior are not comparable to the compound adjectives that eventually form CNs. (Specifically, the numerically prefixed type seems to combine a predicate with an NP rather than a simple N, and the adjectives remain predicating even after the predicate has been deleted.) These adjectives are, however, discussed in more detail in Levi 1975, Section 9.2.

of compound adjective are not recoverably deletable and so may not be included in the RDP set.

Despite the fact that distinct transformations are needed to generate each of the existing types of [predicate + noun] adjectives, the input and output of each of these transformations have enough in common that an adequate grammatical description must somehow reflect these shared features. The principal similarities are (*a*) that the input consists of a predicate element (either participle, preposition, or transitive adjective) followed by a single unmodified noun, and (*b*) that the output shows these two elements combined into a compound adjective which acts syntactically as a single word. In addition, verbs and adjectives regularly follow the nominal element in the compound adjective, while prepositions regularly precede it.

Such structural regularities can be recognized in a grammatical description by characterizing the set of transformations needed to produce these compound adjectives as a **syntactic conspiracy,** and the compound-adjective form itself as a **syntactic target structure** (in the sense of Binnick 1970:241). The striking variety of compound adjectives possible in English, which extends well beyond the few cases of relevance here, is additional evidence for the legitimacy of viewing the compound adjective as a syntactic target structure in English.[24]

The tree diagrams from Table 4.2 which are relevant to Compound Adjective Formation are reproduced in (4.40)–(4.42), together with the necessary conditions that must be imposed on each of the three transformations. Three general comments are in order here to assist the reader in correctly interpreting these diagrams. First, the reader will note that the inclusion of conditions that refer to the source of individual nodes (in these cases, of case markers and of participial suffixes) attests to the globality of these rules. Second, the fact that these trees all present a relative clause configuration within which the rules apply (rather than a simple S) is a function of the fact that the fuller derivations from which the trees are taken are derivations of CNs, which **are** derived from antecedent NPs like those shown here; the relative clause configuration is **not,** however, required for the rules of Compound Adjective Formation to apply. Third, although the trees are drawn to show the underlying VSO order which is assumed here, the examples shown in linear form have been changed into SVO order to increase readability.

[24] A representative selection of the many possible types may be seen in these forms (chosen from the first page of the "lexical index" provided in Meys 1975:222ff.): *age-old, also-ran, animal-loving, anti-hunger, architect-designed, aroma-sealed, atom-free, bare-kneed, behind-the-scenes, birdlike, black-and-white, black–white, blue-eyed, blue–green, broken-down, build-it-yourself, by-passed, clock-wise, cross-dyed.*

(4.40)   Compound Adjective Formation, Type (a):
Passive Participle Type

Example: *stress ## stress be caused by heat ##* ⇒
stress ## stress be heat-caused ##

Input tree:

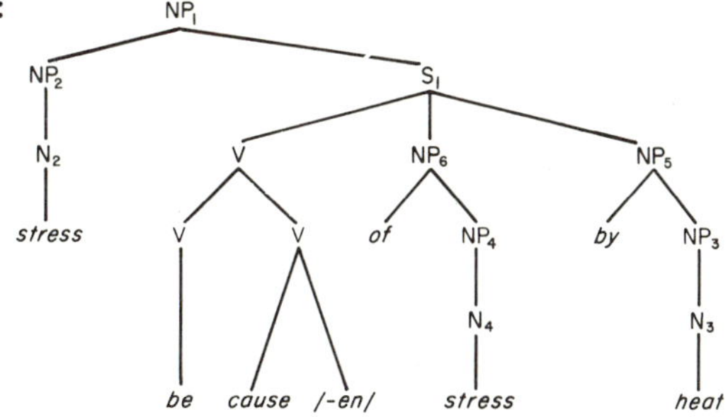

Output tree:

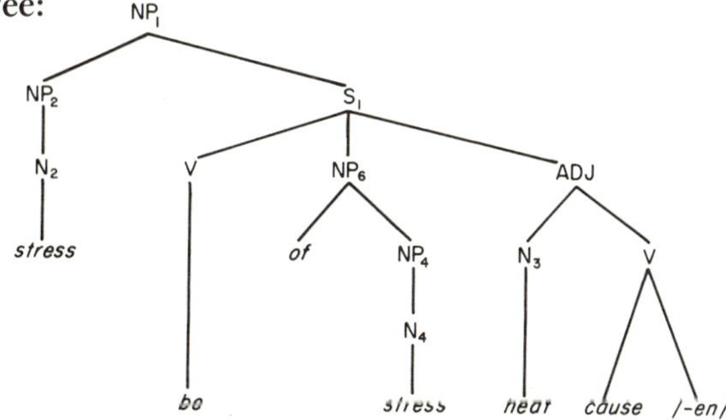

Conditions

1. NP_3 must dominate only an unmodified N.
2. N_3 must be a proper or generic common N.
3. *Of* and *by* (dominated by NP_6 and NP_5, respectively) must be case markers.

Comments

1. /-en/ represents the passive participle suffix.
2. The absence of NP_5, NP_3, and the case marker *by* in the output tree is due to the operation of structure-deleting conventions which remove nodes that no longer dominate terminal elements and a case-deleting convention which removes stranded case markers.
3. This rule applies optionally.

(4.41)   Compound Adjective Formation, Type (b):
Active Participle Type

Example: *clock ## clock make music ## ⇾*
clock ## clock [be] music-making ##

Input tree:

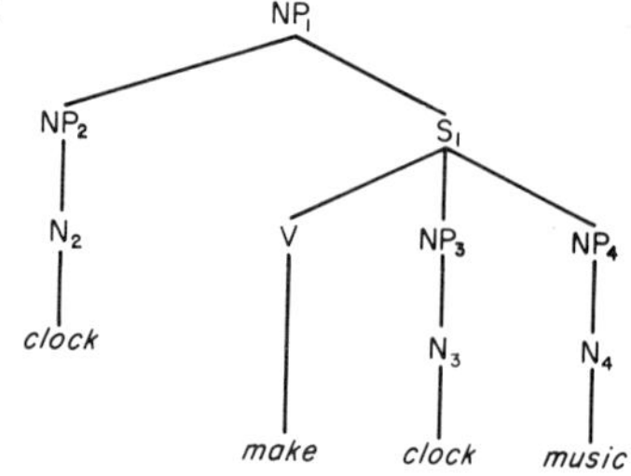

Output tree:

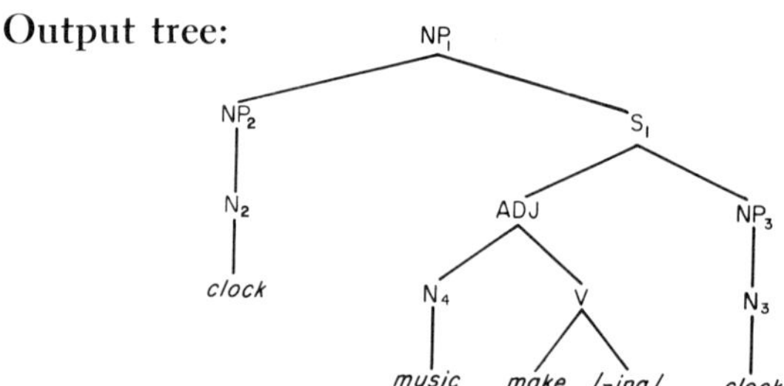

Conditions
1. NP_4 must dominate only an unmodified N.
2. N_4 must be a proper or generic common N.

Comments
1. The participial suffix /-ing/, though shown here, is more probably added by a subsequent rule that operates on unmarked verb stems dominated by ADJ nodes.
2. The nonbranching node NP_4 is presumably removed by a structure-deleting convention accompanying Compound Adjective Formation.
3. Copula insertion may or may not take place subsequently; see discussion earlier in this section.
4. This rule applies optionally.

(4.42)   Compound Adjective Formation, Type (c): Prepositional Type

Example: *note ## note be in margin ## ⇒*
note ## note be in-margin ##

Input tree:

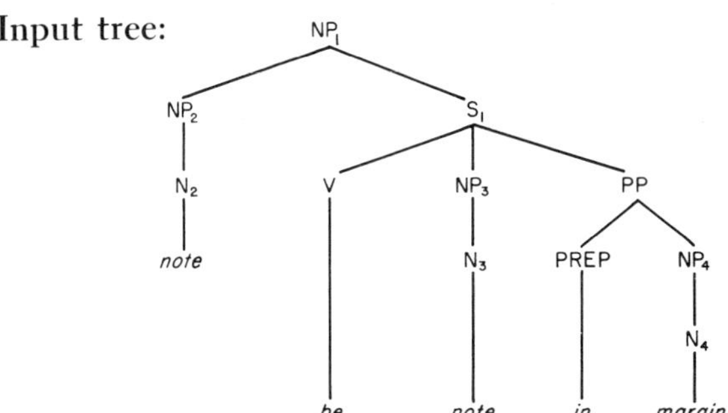

Output tree:

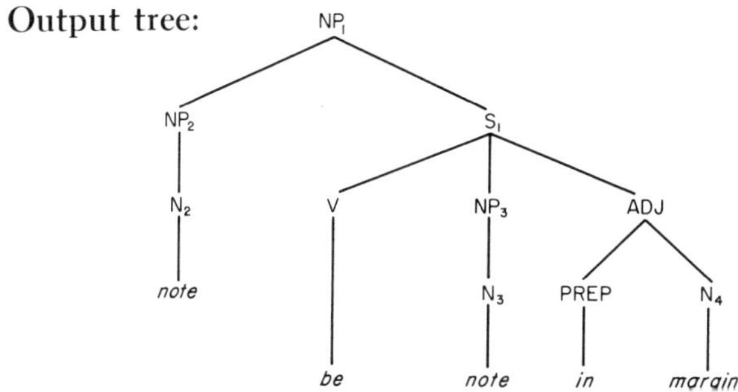

Conditions
1. NP$_4$ must dominate only an unmodified N.
2. N$_4$ must be a proper or generic common N.

Comments
1. Compound adjectives formed by this rule almost always have their prepositional constituent deleted before SS; see later discussion.
2. It is assumed here that the ADJ of the output tree is a new constituent, and hence that the nodes PP and NP$_4$ of the input tree are deleted by a structure-deleting convention accompanying the formation of this ADJ.
3. This rule applies optionally.

The reader will have observed that in each of the preceding input trees, the noun that is to be compounded with the predicate element must be a single, unmodified noun.[25] The purpose of such a condition is to block the eventual formation of such expressions as *a narrow-marginal note*, *a dissonant-musical clock*, or *unusually-high-thermal stress* from the perfectly well-formed underlying structures of *a note which is in a narrow margin*, *a clock which makes dissonant music*, or *stress caused by unusually high heat*, respectively. If we limit the nominal input to unmodified nouns, we shall automatically and appropriately bar the incorporation of more complex forms (i.e., full NPs) into a compound adjective construction. (Similarly, this limitation will prevent the formation of the ungrammatical forms provided in [3.26] of the last chapter.)[26]

It is, however, essential to note that this condition, imposed on all three rules of (4.40)–(4.42), will be satisfied by any CN formed by an earlier application of the CN formation rules since, as has been argued earlier (Section 3.4), the result of that process is a constituent dominated by the node N; it is this constituent structure that contributes to the recursiveness of the CN formation process. Thus, the CN *thermal stress* whose derivation is illustrated in Table 4.1 could itself be input to the Compound Adjective Formation rule, to produce a form such as [*thermal stress*]-*caused*, *in*-[*thermal stress*], and so forth. (Such adjectives might then underlie the modifiers in CNs like [*thermal stress*] *cracks* or perhaps [*thermal stress*] *variations*.)

Another condition on the formation of these compound adjectives is that the noun to be compounded must be either a proper noun, or a

---

[25] The requirement that only unmodified nouns enter into compound formations (as in N–Adj forms like *brick red*, *puzzle-solving*, *student-led*, and *child-free*, as well as N–V forms like *girl-watch* or *hand-deliver*) has been noted before in the literature; for example, Postal (1972:66ff.) discusses its significance for both pseudo-adjective formation in English and object incorporation in Mohawk, while Mardirussian (1975:386) proposes it as a **universal** principle of all noun-into-predicate incorporation processes.

[26] On the other hand, the rules in (4.40)–(4.42) appear unable to account for the formation of technical terms such as these: *full entry theory* (Jackendoff 1975:643), *fast breeder reactor*, *high energy physics*, *dry cell batteries*, *narrow band spectra*, *Small Business Administration*, since the condition just discussed would predict neither their makeup nor the preposability (presumably, as single constituents) of the complex modifiers. However, an explanation may lie in the fact that the complex modifiers are themselves technical terms (an expression intended in a pretheoretical sense here), and hence lexical items in their own right; we may then be justified in considering them to be single nouns, orthography and surface morphology notwithstanding. Although this analysis will not be pursued here, this approach receives support from the fact that structurally parallel phrases whose modifiers do not specify technically defined entities are unacceptable, as predicted: *expensive-cell batteries*, *noisy-band disturbances*, *unprofitable-business subsidies*.

    4 CN DERIVATIONS BY PREDICATE DELETION

generic common noun.[27] This condition is intended to account for the fact that compound adjectives may incorporate specific, referential nouns only when those nouns are proper nouns, as in *Nixon-hating journalists*, *China-watching diplomats*, or *Monopoly-playing teenagers*; when, however, a noun in a compound adjective is a common noun, it may only be interpreted in a generic sense. Thus, a *woman-hating editor* could only describe an editor who hated women in general, not one who hated (atypically) one particular woman while liking many (or even most) others. Similarly, the expression *book-loving children* can only be used to describe children who love books in general but not for children who might have **atypically** formed an attachment to a single book while despising almost all others.

The fact that the common nouns in these compound adjectives are always derived from **generic** antecedents is partially obscured in certain cases by two factors. First, there are cases where extralinguistic knowledge either permits or compels us to interpret a generic noun in a specific, referential sense. For example, *solar generator* would be paraphrased by most people as *generator using [the energy of] the sun*, where *the sun* is a referential, singular noun; I claim, however, that this is the result of our knowledge of the solar system in which we live, and that this reading is superimposed on the basic generic reading. Thus, if it should prove technically feasible to reach the suns of solar systems beyond our own so that we might tap a number of suns for their energy, *solar generator* could then be regularly interpreted as *generator using [the energy of] suns*, that is, *suns* in the generic sense.

The second factor that tends to obscure this requirement of genericness is that only one member of a class may be relevant in a particular context, even though the CN strictly speaking has a generic interpretation. Thus, a friend who is a rabbi could be called a *rabbinical friend*, because he or she would be a member of the set described by *friends who are rabbis*; nonetheless, the CN *rabbinical friend* could not be derived from *a friend who is the rabbi*, where *the rabbi* would refer to a specific person or specific position in a community. Similarly, a seminar or-

---

[27] Although these two types of nouns appear at first glance to have little in common, a deeper analysis reveals that they both represent ways of exhaustively enumerating a set of objects. Proper nouns exhaustively enumerate one-member sets simply by naming the unique member itself; generic nouns, on the other hand, exhaustively enumerate a set essentially by using the predicate nominal that characterizes all its members. The union of these two groups of nouns in the structural description of these rules is therefore not at all as unmotivated as it first appears. (Postal [1969:233] makes a similar observation in discussing the distribution of the expression *so-called*, which also must be described in terms of the union of these two noun classes.)

ganized by one student is necessarily a *student-organized seminar* because it is one instance of the more general type (i.e., *seminars organized by students*); what is important to recognize here, however, is that the term is applicable only by virtue of its derivation from the more general type and not because it is a legitimate term for the specific instance alone.

The demonstrably recursive nature of CN formation implies that the rules shown in (4.40)–(4.42) must be considered cyclic rules, since the entire process may be repeated any number of times with the output of one cycle constituting the input to the next. (For example, the cyclic application of (4.42) together with other relevant rules can generate not only *doghouse*, but subsequently *doghouse roof, doghouse roof nails, doghouse roof nail barrel, doghouse roof nail barrel staves*, and so forth, by repeated compounding of the previous CN with the preposition *for*, followed by deletion of the latter as an RDP.)

It is interesting to observe that the demonstrable recursiveness of the rules of (4.40)–(4.42) is virtually the only evidence available in support of treating these rules as cyclic ones.[28] If we try to construct the usual kind of cyclicity argument based on interdependencies between these rules and other syntactic rules already known to be in the cycle, we find almost no observable interactions at all. The one case where an argument based on intrinsic ordering can be made involves the cyclic rule of Passive, since that rule must be applied before the rule of (4.40), at least, in order that compound adjectives based on passive participles (e.g., *heat-caused*,

---

[28] Newmeyer (1975) argues that compound formation, along with predicate raising and nominalization, is actually a precyclic process. I believe, however, that his argumentation depends crucially on a number of untenable claims which can only be briefly indicated here. First, his proposals concerning precyclic compound formation are advanced primarily to preserve an earlier claim (Newmeyer 1976, originally available in 1974) that nominalization is also precyclic; however, the evidence in Section 5.4 to follow will show that nominalization must be regarded as a cyclic process. Second, the "nonrelatival" noun complement source that he sets up for CNs analyzed here under RDP **prepositions** in particular (such that *tree house* is derived directly from *house in tree* rather than from *house which is in tree*) has three serious drawbacks: (*a*) It rests on arguments that are primarily weak or just untenable (together with some more valid points which are, however, captured in the present analysis); (*b*) it creates a derivational dichotomy among the RDP-type CNs (between those with prepositional RDPs and those with verbal ones) of a much more radical nature than the evidence warrants; and (*c*) it sacrifices a number of theoretical advantages inherent in providing a sentential source for all these CNs since it would require duplicated statements of semantic co-occurrence restrictions and aspectual constraints, and would preclude possibilities for case marking CNs with underlying RDP verbs. Finally, Newmeyer's proposal would have the result of deriving some CNs precyclically and others cyclically, when in fact the syntactic interaction of all CN types (those with RDP prepositions, those with RDP verbs, and NOM CN types) requires a unitary derivational analysis which is, moreover, demonstrably a cyclic one. (See Section 5.4 for further discussion.)

government-sponsored, student-run) may be formed. Since, however, this ordering relationship only tells us that (4.40) must apply **after** Passive, this fact is evidence against the **pre**cyclicity of (4.40) but insufficient evidence, in and of itself, for its cyclicity (since a postcyclic application would also be post-Passive). Moreover, a comparison of the rules of (4.41) and (4.42) with known cyclic rules turns up no evidence at all for their cyclicity that would be based on ordering relationships among these rules.

Whether we can find a principled explanation for this somewhat surprising lack of interaction between the syntactic rules of the cycle and the rules involved in CN formation is a question that has no satisfying answer at present. Although it is possible that an answer may be found in distinguishing, in some more systematic way than has been attempted here, between word formation rules on the one hand, and syntactic or "sentence formation" rules on the other, such a conjecture cannot be pursued within the bounds of the present study. Instead, the rules of (4.40)–(4.42), as well as other crucial rules involved in the formation of CNs, will be treated as cyclic rules based on the still telling argument of their indisputable recursive potential.

In addition to being cyclic, these rules must all be classified as optional transformations since the relative clauses that constitute their inputs are perfectly grammatical without further transformation; thus, *stress which is caused by heat, clocks which produce music,* and *notes which are in margins* are all just as acceptable as their ultimate descendants *thermal stress, musical clocks,* and *marginal notes*. It is not, however, the case that the **outputs** of these transformations are all equally grammatical; for example, virtually all compound adjectives whose predicates are RDPs in prepositional form must eventually undergo RDP Deletion. It is also true that most adjectives formed with an active participle are grammatical only in prenominal position (as in *woman-hating editors* or *home-owning blacks*) but not when left in predicate position (as in **editors who are woman-hating* or **blacks who are home-owning*), even though the latter configuration is fine with related agentives (such as *editors who are woman-haters* or *blacks who are home-owners*).

Other compound adjectives that are never lexicalized per se but must always undergo subsequent transformations include those that are based on any participles formed from BE and HAVE. Thus, the relative clause construction of *friends who are professors* apparently must go through two ungrammatical intermediate stages of **friends who are professor-being* and **professor-being friends* before having the participial predicate obligatorily deleted to form *professor friends* (cf. *student friends*) or *professorial friends*.

In addition, neither of the two participles that may be formed from

HAVE are ever lexicalized per se in compound adjectives (although the active participle does surface in reduced relatives [e.g., *adolescents having problems, a profession having its own jargon*]). Thus, the intermediate stages marked with asterisks in (4.43) never reach the surface without further change.

(4.43)

 a. *atlas which has pictures*   b. *intuition which women have*
  **atlas which is picture-having*   **intuition which is had by*
  **picture-having atlas*   *women*
              **intuition which is women-had*
              **women-had intuition*

Although one may have doubts as to whether the compound adjectives shown in (4.43) are ever generated at all, this question will not be pursued here; instead, we may observe that the relative clauses seen in (4.43) are in fact regularly replaced in English by suppletive prepositional phrases in *with* (corresponding to the active participle *having*), or the genitive *of* (corresponding to the passive participle *had*). These postnominal expressions may then be transformed into prenominal modifiers, as shown by the fuller derivations sketched in (4.44); in such an analysis, the compound adjectives for HAVE-type CNs would be formed by the rule of (4.42) rather than the rules of (4.40)–(4.41). Note that the genitive *of* in (4.44b) sometimes survives as the suffix *'s* in prenominal position, whereas the compound adjective formed from *with* deletes its RDP preposition with no remaining trace.

(4.44)   a. *atlas which has pictures*   b. *intuition which women have*
      atlas with pictures   *intuition of women*
      **with-picture atlas*   **of-women intuition*
      picture atlas   *women's intuition*
      pictorial atlas   *feminine intuition*

Although the intermediate stages marked with asterisks in (4.44) may strike us as unnatural or weird, this "weirdness" is typical of compound adjectives formed with prepositions that belong to the RDP set; thus, *marital sex* and *avian sanctuaries* have equally weird intermediate stages of **in-marriage sex* and **for-bird(s) sanctuaries*. This intermediate stage is nonetheless justified by the many compound adjectives that do surface with their prepositions intact, as in *between-meal snacks, offshore drilling, on-site inspections, after-hours liaisons*, and the many adjectives whose underlying prepositions surface as prefixes instead, as in *prewar prices, intramural games, postnasal drip*, and *transoceanic shipping*. The weirdness of **with-picture atlas* is therefore not directly attributable to the Preposition + Noun configuration, since that is a common one in English. Instead it must be explained by the fact that we are strongly aware that those prepositions that are derived from RDPs (i.e., *in*, *at*, *on*, *for*, *from*, *about*, and the prepositional descendants of HAVE participles) may not surface in compound adjectives but must be transformationally deleted before reaching the surface.

4.2.3 Recoverable Predicate (RDP) Deletion

Recoverable Predicate Deletion is the process that (*a*) deletes from a prenominal compound adjective that predicate element, whether participle or preposition, which is a lexicalization of any member of the RDP set, and (*b*) Chomsky-adjoins the remaining noun to the head noun of the NP. Since this adjunction of the two nouns under a new N node is the step which actually creates the inviolable syntactic unit of the CN, this process might also be called "Complex Nominal Formation." (There are, however, a number of different rules that create CNs, including the rules described in this section and the rules of Object Preposing and Subject Adjunction to be analyzed in Chapter 5.) Generalized descriptions of the rules which delete RDPs lexicalized as verbs and the rules which delete RDPs lexicalized as prepositions are shown in (4.45) and (4.46), respectively.

(4.45)   Recoverable Predicate Deletion, Verbal Type

Example: $heat \begin{Bmatrix} \textit{-caused} \\ \textit{-causing} \end{Bmatrix} stress > heat\ stress$

Input tree:

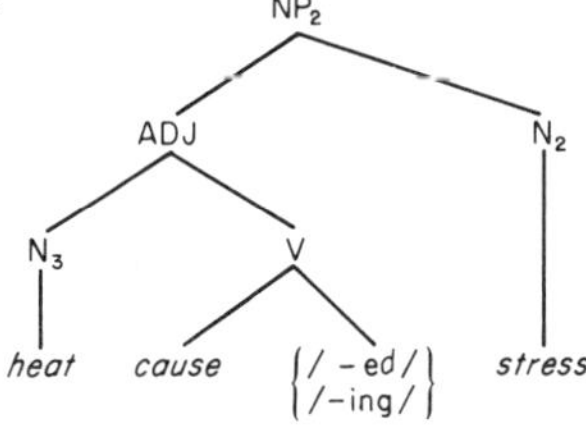

Output tree:

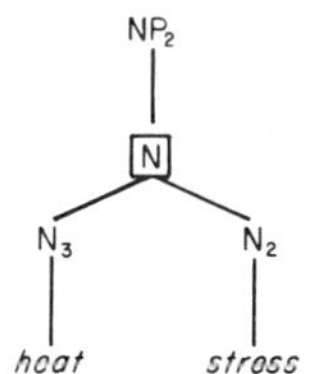

Condition

The V node of the compound adjective must dominate a lexicalization of a member of the RDP set.

Comment

1. As indicated in the input tree, the verb stem may have either the past participle suffix shown here as /-en/ or the active participle suffix /-ing/ as a right sister; RDP Deletion operates in the same manner in both cases.
2. The N node shown within a box is the node created by the Chomsky-adjunction of N_3 and N_2; it marks the first appearance in the derivation of the complex nominal per se.
3. This rule applies obligatorily; for the few exceptions, see discussion to follow.

(4.46)   Recoverable Predicate Deletion, Prepositional Type

Example: *in-margin note > margin note*

Input tree:

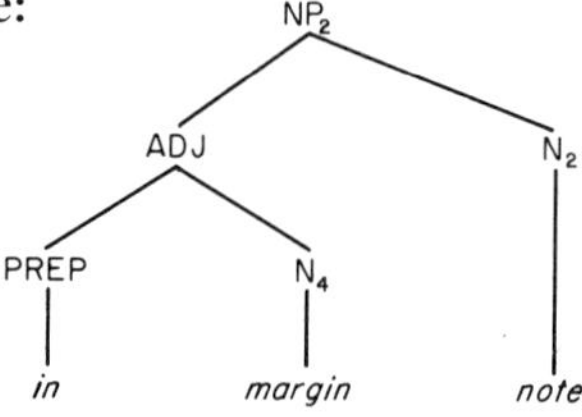

Output tree:

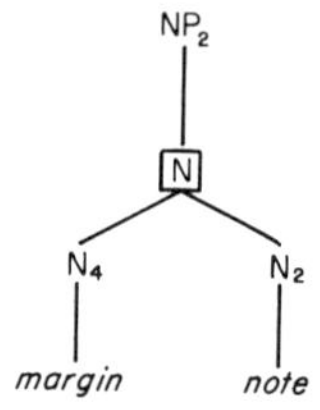

Condition

The PREP node of the compound adjective must dominate a lexicalization of a member of the RDP set.

Comments

1. The N node shown within a box is the node created by Chomsky-adjunction of N_4 and N_2; it marks the first appearance in the derivation of the complex nominal per se.
2. This rule applies obligatorily; for the few exceptions, see discussion to follow.

Although (4.45)–(4.46) describe the two major types of RDP Deletion rules, we must recognize that this description is really an abbreviation for a total of nine distinct rules—one for each of the RDPs—which happen to group themselves into the two patterns shown in (4.45)–(4.46). No single rule of RDP Deletion can be written, not only because of the two different constituent orders observable within the compound adjectives (i.e., predicate–noun versus noun–predicate), but, more importantly, because there is no single semantic or syntactic feature that unites the RDPs in such a way that they could be characterized as a **class** (in a noncircular fashion) within a single transformation. We may therefore speak of the RDP Deletion **process,** but of the RDP Deletion **rules** [plural]. Once again in this study of CNs, we observe a conspiracy of separate rules (this time nine in number) operating on somewhat different inputs to produce a single target structure: the complex nominal. What makes this particular syntactic conspiracy more striking than some others is the fact that the "conspirator rules" are remarkably similar in both input and output, the only differences being the specific RDP named in each case, and the constituent order within the compound adjective involved.

Any rewriting of the generalized formulations of (4.45)–(4.46) into their nine component rules would necessarily have each RDP specified in its **semantic** form within the Condition imposed on each rule (e.g., "Condition: The V node of the compound adjective must dominate a lexicalization of the predicate CAUSE" or "Condition: The PREP node of the compound adjective must dominate a lexicalization of the predicate IN"). This is required since each deletion rule operates only on those lexical items that are direct "descendants" of specific predicates in the RDP set; homonymous forms from different sources (such as a case marker *in*, a 'favoring' *for*, or an 'approximate' *about*) would be ignored by these rules. Because of this modus operandi, however, we must recognize that the RDP Deletion rules are also global rules.

Let us now take another look at the change in constituent structure effected by these rules, as indicated here:

(4.47)

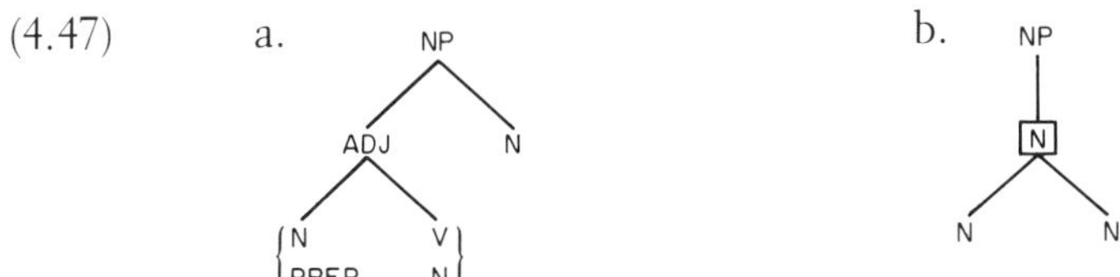

The reader can easily see from these diagrams that the configuration on the right represents the first appearance in the derivation of that distinctive structure which characterizes the complex nominal: namely, a noun node dominating two daughter nouns (of which the left may subsequently be adjectivalized).

The most important difference in constituent structure between the two configurations is that in the first, the head noun is modified by an adjective (here, of compound form) which is a **separate** constituent, whereas in the second, the modifying element that has survived RDP Deletion (i.e., the noun from the compound adjective) has become an integral and **inseparable** part of a new head noun, created by Chomsky-adjunction and dominating the old head noun. This new head noun is, of course, the complex nominal itself.

It is this change in constituent structure that explains the observation made much earlier that nominal adjectives may not be separated from their head nouns by any intervening material; the nominal adjective (like the modifying noun that is its immediate ancestor in the derivation) and the original head noun together constitute a single noun, and there is simply no way that this new noun may be breached.[29] In contrast, compound adjectives, just like other predicating adjectives, form constituents that are separate from their head nouns (as indicated in [4.47a]) and so permit the intercalation of other lexical material between them and the head noun. This difference is illustrated in (4.48), where the inseparability of the nominal adjectives in (a) from their head nouns is contrasted with the syntactic autonomy of the predicating compound adjectives in (b).

(4.48) a. *governmental (*erroneous) decisions*
*an electric (*and expensive) shaver*
*an occupational (*needless) hazard*
*viral (*and virulent) diseases*

   b. *her prizewinning (irresistible) recipes*
a sun-drenched (peaceful little) retreat
oil-covered (and thoroughly wretched) birds
several meaning-bearing (but ambiguous) morphemes
an underwater (and highly dangerous) mission

The reader will also recall that nominal adjectives in CNs were originally distinguished from normal adjectives by the fact that only the latter could

---

[29] The only way in which the two constituent parts of a complex nominal may be separated from each other is in coordinate structures like *civil and mechanical engineering, electric and manual typewriters, gas and solar heating;* since, however, this separability within coordination may also be observed between bound morphemes and stems which otherwise must be considered to form single words (as in *pre- and postnatal care, pro- and antibusing forces, micro- and macroeconomics, psycho- and sociolinguistics*), this separability may not be counted as evidence against the claim that each CN is indeed a single noun.

   4 CN DERIVATIONS BY PREDICATE DELETION

appear freely in predicate position. We now see that the two types of adjectives are also systematically distinguished in the internal structure of surface NPs; specifically, predicating adjectives (in prenominal position) are immediately dominated by the higher NP node, whereas nominal adjectives are separated from that NP node by the N node immediately dominating the entire CN construction.

Although in general the RDP rules appear to apply obligatorily, there are a few exceptions; specifically, the RDP rules which specify CAUSE, MAKE, and IN sometimes permit the full compound adjectives to survive to Surface Structure (though the latter two do this very rarely indeed). Data illustrating this distinction are given in (4.49).

(4.49)

RDP	Data supporting obligatory deletion	Data supporting optional deletion
CAUSE		*heat-caused stress* *virus-caused pneumonia* *malaria-causing mosquitoes*
HAVE	$\begin{Bmatrix} \text{*picture-having} \\ \text{*with-picture} \end{Bmatrix}$ *atlas* $\begin{Bmatrix} \text{*women-had} \\ \text{*of-women} \end{Bmatrix}$ *intuition*	
MAKE	**consonant-made patterns* **flower-made wreaths*	*candy-making factory* *noise-making instruments* *trouble-making kids*[30]
USE	**steam-using iron* **sun-using generator*	
BE	**rabbi-being friends* **mammal-being vertebrates*	
IN	**in-marriage sex* **in-summer travel*	*in-service training* *in-house editing* *?in-town dining*
FOR	**for-mayor salary* **for-nose spray*	
FROM	**from-olive oil* **from-the-store teeth*	
ABOUT	**on-tax law* **about-history lecture*	

[30] The fact that *trouble-making kids* may not be reduced to *trouble kids* is probably due to the fact that the participle here has the sense of 'cause' rather than the sense of 'physically produce'. It therefore belongs to the N-*causing* group of "pre-CN expressions" which show an inconsistent pattern of RDP Deletion; this particular example thus works like *cancer-causing chemicals*, cited earlier, which does not permit CAUSE Deletion, and unlike *disease-causing germs*, which does.

The fact that the net effect of the nine RDP Deletion rules is to create a potential 12-way ambiguity (reflected in the 12 groups of CNs shown earlier in [4.1]) for any given CN brings us now to the issue of "recoverability conditions" on deletion transformations. Although the overall theoretical need for such constraints (and/or their appropriate formulation) will not be settled here, it may nonetheless be of some interest to consider briefly how the present analysis fits in both with classic formulations and with more recent approaches.

The classic statement is to be found in Chomsky 1965:

> We are proposing the following convention to guarantee recoverability of deletion: a deletion operation can eliminate only a dummy element, or a formative explicitly mentioned in the structure index (for example, *you* in imperatives), or the designated representative of a category . . . , or an element that is otherwise represented in the sentence in a fixed position [pp. 144–145].

In one sense, each of the nine RDP Deletion rules meets this constraint since the RDP itself may be viewed as an "explicitly mentioned formative"; we must note, however, that the deletable item is defined in semantic rather than lexical terms, an approach that we can be sure Chomsky did not have in mind.

The adequacy of Chomsky's formulation, however, has been questioned in more recent years in analyses by McCawley (1973, 1976b) and Hankamer (1973), not so much on the grounds that such a constraint is unnecessary in a grammar, but more on the grounds that (*a*) the notion of "recoverability" itself has yet to be adequately defined or justified, and (*b*) the set of deletion processes in natural languages is more heterogeneous than had previously been recognized, so that constraints proposed for one kind of deletion process may be inappropriate for another. In the present context, for example, the fact that the constraints on RDP Deletion appear fundamentally different from constraints that have been proposed for such rules as VP Deletion and Gapping may be due to the fact that only the former are involved in the process of **word formation,** an area of the grammar that shares some but by no means all of the features of what we might call purely syntactic processes. (Another hypothesis we might consider is that rules that restructure NPs may be subject to qualitatively different constraints than rules that restructure Ss.)[31]

What can we learn about recoverability of deleted material from our present understanding of CNs and the RDP Deletion rules proposed here? To begin with, we certainly must recognize that despite the poten-

---

[31] Although this hypothesis is not explored directly in this book, just such an analysis has been advanced in recent years in the form of lexicalist proposals for the inclusion of an NP cycle within syntactic theory. For details, see Akmajian 1975 and Williams 1974.

                4 CN DERIVATIONS BY PREDICATE DELETION

tial multiple ambiguity set up by the RDP Deletion rules, CN forms **are** used regularly, frequently, and **easily** in a wide variety of natural languages; the problem of "recovering" the deleted material appears no more serious for CNs than for any other regular deletion process that also sets up ambiguity. This would indicate, for one thing, that any constraint as powerful as the No-Ambiguity Condition (NAC) proposed by Hankamer (1973) for deletion over variables would be much too strong to apply to processes involved in CN formation.

It may be that we need to pay closer attention to the notion of **target structure** as we try to achieve a better understanding of the function and limitations of various deletion processes. One hypothesis to consider, for example, is that certain syntactic constructions are so highly functional in a language (in terms of such factors as linguistic conciseness and/or ease of processing) that the constraints on the deletion processes that produce these configurations will be relatively weak, in order that a relatively large number of source structures may converge into such "target structures"; examples of such highly exploited configurations would include the N–N and Adj–N varieties of CNs, as well as the V–NP–NP structure discussed in Green 1974. On the other hand, if other syntactic structures are for some reason less useful or less suitable for "targeting" in the language, then the deletion processes producing them might be subject to stronger constraints since no functional purpose would be served in relaxing them; an example of this sort might be the rather odd surface structures produced by Gapping rules. **If** such a contrast makes sense, then we would expect that constraints proposed for rules like Gapping would be inappropriate for, and perhaps even irrelevant to, the kinds of rules required for processes like CN formation.

Satisfying answers to these questions are not available at present, although it is hoped that increased attention to word formation processes like CN formation will result in the development of a broader and more adequate theory of deletion processes in general. In addition, the recent tendency of linguists to devote more attention to the communicative **function** served by syntactic constructions which were previously studied in fastidiously idealized and context-free form cannot help but facilitate the search for such a theory.

4.2.4 Morphological Adjectivalization (Morph Adj)

This is the transformation that is responsible for producing nominal adjectives from prenominal nouns in CNs, as in *marginal note* from *margin note*, *thermal stress* from *heat stress*, and *musical clock* from *music clock*. The change effected by this transformation is a highly superficial one; as noted earlier, the semantic and syntactic characteristics of CNs are the same, regardless of whether the prenominal modifier is morphologically a noun or an adjective. Indeed, there exist numerous CN pairs such as *atom bomb* and *atomic bomb* which are synonymous, differing in their derivations solely by the application or nonapplication of this rule. This rule should therefore be regarded, as its name suggests, as a fundamentally morphological rule (in the sense that it adjusts superficial morphology) rather than as a syntactic rule. The change it produces is sketched in (4.50), in which x is the lexical noun inserted early in the derivation, x′ is the adjective listed in the lexicon as the nonpredicating counterpart of x, and Y is whatever other prenominal elements may be permitted within that NP.

(4.50)

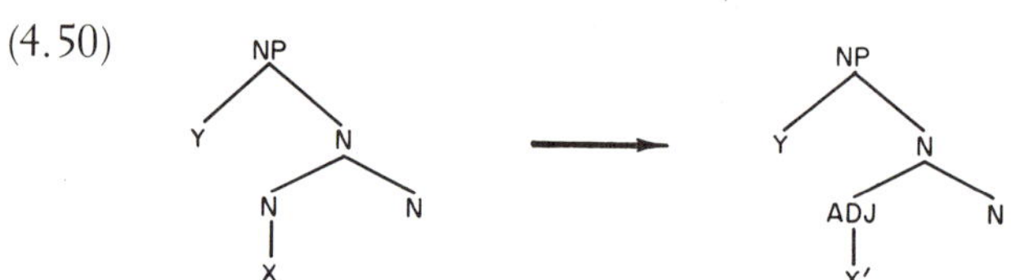

The immediate domination of the nominal adjective by N rather than NP is a syntactic reflection of its nominal (and hence nonpredicating) origins, since the only adjectives to appear in such a configuration are those derived from immediately antecedent nouns within CNs. In addition, this structure shows that the adjective forms a single constituent with its right sister (the original head noun), from which it may not be separated; this pattern too serves to distinguish the nominal adjective from nonnominal counterparts.

As suggested by these diagrams, the Morph Adj transformation must be formulated in such a way as to adjectivalize only those nouns which constitute the **left** branch of any of the binary adjunctions that may be contained within a CN. Thus, if we consider the small sample of possible CN configurations shown in (4.51), we see that this "left-branch condition" will permit only those N nodes appearing in boldface to be adjectivalized by this rule.

(4.51)

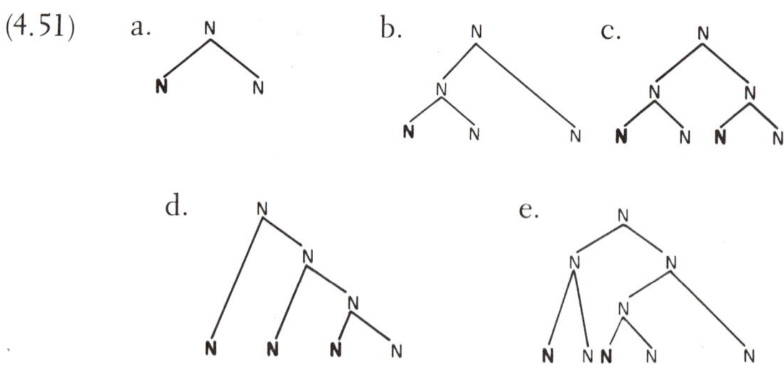

          4 CN DERIVATIONS BY PREDICATE DELETION

Specific CNs corresponding to three of the more interesting structures of (4.51) are provided in (4.52)–(4.54), with brackets indicating the appropriate constituent structure. In each case, the form in (a) shows the CN in its basic N–N form, prior to the application of Morph Adj; the form in (b) then shows the output of this rule when applied correctly, while the form in (c) shows the ungrammatical result of applying Morph Adj to nodes that are not on left branches of their CNs.[32]

(4.52)   a. [*moon city*] *entertainment*
           b. [*lunar city*] *entertainment*
           c. *[*moon urban*] *entertainment*

(4.53)   a. [*crime law*] [*instruction materials*]
           b. [*criminal law*] [*instructional materials*]
           c. *[*criminal legal*] [*instructional materials*]

(4.54)   a. [*industry society*] [*automobile mechanics*] *manpower*
           b. [*industrial society*] [*automotive mechanics*] *manpower*
           c. *[*industrial social*] [*automotive mechanical*] *manpower*

Since this left-branch condition is certainly not a priori necessary, we might well ask what function it serves in the language. I believe the answer is a pragmatic one, relating to ease of processing recursively formed CNs, and based on the fact that an extension of Morph Adj to apply to virtually any N within a complex CN would obliterate the internal structure of that CN and make it difficult if not impossible to process. The reader will note, for example, that the adjectives in the ill-formed (c) examples just considered have the effect as we read them of shifting the bracketing, so to speak, in our minds; instead of the bracketing shown here for each CN, we mentally insert new brackets before every nominal adjective, and, in so doing, destroy the original constituent structure of the sources indicated in (a). To take the problem one step further, suppose that Morph Adj were to apply to every noun but the last in this very complex but still understandable CN: *industrial society recreational department financial manager psychological profile*; the result

---

[32] The only cases where this condition may be violated with acceptable results are found in a small set of highly technical terms such as [*set theoret*]*ical analysis*, [*ego psycholog*]*ical framework*, [*quantum mechanic*]*al interpretation*, or [*text grammat*]*ical model*, where the brackets are intended to show that the suffix has been added to an entire CN. Note, however, that the strength of this left-branch condition is actually attested to here in that (*a*) no more than one violation ever occurs in a single CN; (*b*) speakers are not free to coin additional examples; and (*c*) these forms are quite rare even within their own very specialized disciplines. (The fact that each CN shown here within brackets is itself a technical term, and hence a lexical item in its own right, probably strengthens our sense of the "nouniness" of these forms, and thus facilitates the attachment of a suffix normally used only for noncomplex nouns.)

would be the incomprehensible *industrial social recreational departmental financial managerial psychological profile.*

It thus appears that the left-branch condition on Morph Adj actually facilitates our perception of CNs since every occurrence of a nominal adjective tells us in effect to group it and the following noun into a single constituent; the ease with which we follow this instruction is due to our familiarity with the binary structure characteristic of every CN, as well as with the extremely general interpretive principle that a prenominal adjective in English is a modifier of the following noun. If, however, this very regular pattern were to be obscured as the result of a more indiscriminate application of Morph Adj, we could no longer "recover" the internal bracketing of the CN and would then have significantly greater difficulty in understanding unfamiliar and/or particularly complex constructions. It is for this reason that even within the technical style which permits adjectivalization of right-branch nouns (see footnote 32), this "exemption" is granted very sparingly. To be sure, we may observe a few forms like *truth functional analysis* or even *relational grammatical interpretation*, but we can now understand why we would not expect to find forms like the ones cited earlier where the constraint has been relaxed without limit.

The evidence relating to the ordering of Morph Adj suggests that it applies quite late in the derivation, although it is difficult to say just how late. It certainly must follow the RDP Deletion rules, since its structural description is not met until after the latter have produced the predicateless CN. To illustrate with the paradigmatic forms of this chapter, *heat* will be transformed into *thermal* and *music* into *musical* only after RDP Deletion has applied; this explains the possibility of *thermal stress* and *musical clock* but the impossibility of both **thermal-caused stress* and **musical-making clock.*[33]

The data in (4.55)–(4.57) suggest that Morph Adj should also be ordered before *one*-pronominalization, since the application of the latter appears to require the prior application of the former.

(4.55) a. *Those speeches are all* $\left\{\begin{array}{l}presidents'\\presidential\end{array}\right\}$ *speeches.*

b. *Those speeches are all* $\left\{\begin{array}{l}*president(s')\\presidential\end{array}\right\}$ *ones.*

---

[33] In some cases, compound adjectives composed of nouns and passive participles may undergo a different rule, which changes certain kinds of nouns into superficial adverbs, as in *chemically caused hallucinations* or *electrically powered engines*. These forms do not concern us here, however, since a form which undergoes this rule will be effectively removed from the derivational path that leads to CN formation.

(4.56) a. *Those analyses are all* $\left\{\begin{matrix}microscope\\ microscopic\end{matrix}\right\}$ *analyses.*

  b. *Those analyses are all* $\left\{\begin{matrix}{}^{*}microscope\\ microscopic\end{matrix}\right\}$ *ones.*

(4.57) a. *Those infections are all* $\left\{\begin{matrix}virus\\ viral\end{matrix}\right\}$ *infections.*

  b. *Those infections are all* $\left\{\begin{matrix}{}^{*}virus\\ viral\end{matrix}\right\}$ *ones.*

On the other hand, the distribution of the asterisks in these examples might be better explained in terms of an output constraint that would block the appearance in SS of CNs of the form [N + *one* (s)]. This analysis appears preferable in view of the fact that the constraint we would have to impose on *one*-pronominalization is essentially a negative one (i.e., "Don't apply if sister node is a noun" or "Apply unless there is a left sister labeled N"); in addition, an analysis in terms of an output constraint would fit better with the fact that this constraint is relaxed in some dialects in at least informal speech.

If we choose to explain the data in (4.55)–(4.57) by an output constraint, however, then we are left with virtually no evidence to help us determine just how late the rule of Morph Adj applies. Although a postcyclic application is certainly possible, this will not be argued for explicitly here; instead I will simply treat Morph Adj as a very late rule which is the last to affect directly the derivation of CNs derived by predicate deletion.

In view of the vast number of CNs with an N–N structure as well as the many pairs of synonymous CNs like those shown in (4.58), it is clear that the Morph Adj rule must be an optional one. The question then to consider is, what are the conditions which influence and/or determine its application?

(4.58) *atom bomb*   *atomic bomb*
  industry psychologist   *industrial psychologist*
  ocean winds   *oceanic winds*
  city life   *urban life*
  hand signal   *manual signal*
  sound system   *acoustic system*

To begin with, we must note that the rule may be prevented from successfully replacing a left-branch noun with an appropriate nominal adjective simply because of lexical gaps. Some nouns (most often of

Germanic origin) simply do not have corresponding nominal adjectives in English; this is the case for the prenominal modifiers in CNs like *mining engineer*, *divorce law*, and *harbor policeman*. In other cases, nonpredicating adjectives do exist and thus are readily substituted for the noun; such is the case for the parallel CNs *sanitary* (< *sanitation*) *engineer*, *criminal* (< *crime*) *law*, and *urban* (< *city*) *policeman*. Note that although morphological similarity is often observed between a noun and its nominal adjective, this is neither a necessary nor a sufficient condition for choosing the correct adjectival counterpart to a given noun. That it is not a **necessary** condition is demonstrated by the large number of suppletive adjectives of Romance origin which serve as nominal adjectives for Germanic nouns, as in *oral* for *mouth*, *feline* for *cat*, *solar* for *sun*, and *marine* for *sea*. That morphological similarity between an adjective and a noun is not a **sufficient** condition for substitutability in CNs is demonstrated by the nonequivalence of such pairs as *blood test* and *bloody test*, *mountain resort* and *mountainous resort*, *energy demands* and *energetic demands*. (These particular pairs are nonequivalent because the suffix added to the nominal stem in each case carries semantic meaning of its own, whereas the suffixes transforming nouns into nominal adjectives must be semantically empty.)

Because the pairing of nouns and their corresponding nonpredicating adjectives is not simply a matter of attaching some predictable suffix to a nominal stem, it appears to be necessary that the lexicon be structured in such a way (for languages with a Morph Adj rule) as to include in the lexical entry for each noun a specification of the adjective that may replace it in a CN. Such pairings are indispensable in a complete lexicon in view of the fact that there is no way to tell a priori whether the nominal adjective is produced by a relatively straightforward suffixation pattern (e.g., *music* > *musical*), by a partially idiosyncratic pattern (e.g., *crime* > *criminal*), by a suppletive stem with a normal suffix (e.g., *sun* > *solar*, *mouth* > *oral*), or, for that matter, whether there is any nominal adjective at all even in the most technical levels of discourse. (The absence of an appropriate adjective seems as pertinent a fact for the lexical entry of a given noun as the specification of an existing one.) In a few cases, the information provided by these pairings is needed for the additional reason that the same lexical noun may have a nominal adjective associated with it on one reading, and not on another. Thus, *circulation system* may be replaced by *circulatory system* to describe the biological circulation of blood, but not the circulation of books or motor vehicles.

This pairing of noun and adjective would be comparable to the pairing of verb and noun that is necessary to provide the correct morphological form in nominalizations (e.g., *decide* and *decision*, *remove* and *removal*, *grow* and *growth*, *criticize* and *criticism*). But where the nouns inserted after nominalization rules have applied must incorporate the semantic content of both the verb and a specific head noun (such as /ACT/ or /PRODUCT/), the adjectives selected and then inserted by the Morph Adj rule must contain absolutely no additional semantic content beyond that of the noun itself.

As noted earlier, the etymology of a noun is a major factor in predicting the **existence** of nominal adjectives, since nouns of Romance and Greek origin are provided with a much richer assortment of possible derivational suffixes than are nouns of Germanic origin. (Consider, for example, the enormously productive set of related suffixes *-al*, *-ical*, *-ial*, *-oidal*, *-orial*, and *-ual*, all of which may make adjectives from Romance or Greek nouns but almost never from Germanic nouns.) Etymology is, however, insufficient to explain in all cases the **occurrence** (or nonoccurrence) of such adjectives in specific CNs. The reason for this is that there are sociolinguistic factors of various kinds (including stylistic level and accidents of historical usage) that restrict the application of Morph Adj in ways that are independent of the lexical availability of the appropriate adjective. These sociolinguistic constraints will be discussed in more detail in Section 6.1. For the present, it suffices to state that Morph Adj is a fundamentally optional rule whose successful operation can be blocked in two different ways: either by lexical gaps or by a variety of sociolinguistic factors.

4.3 Summary of Evidence in Support of RDP Deletion

In the preceding sections of this chapter, considerable detail was provided concerning the mechanics of the derivations which produce CNs by predicate deletion. Certainly the major theoretical innovation inherent in these derivations is the claim that there exists a small set of Recoverably Deletable Predicates, which carry important semantic material but which may be deleted by "RDP Deletion" rules before Surface Structure. The present section will be devoted to a summary of the most important pieces of evidence that may be cited in support of this crucial process of RDP Deletion.

The major evidence which supports such a process is, of course, the fact that it permits us to make a vast number of correct predictions about the set of possible semantic structures that can be correctly associated with a given surface CN. Earlier studies of CN forms failed to make such predictions, either because the authors felt, like Jespersen (1942), that "the number of possible logical relations between the two elements is endless [Vol. 6, p. 143]," or because the authors limited their studies to the ambiguities created by different **syntactic** sources (as in Lees 1960) without attempting a detailed analysis of the possible semantic relationships as well. The theory presented in this book thus constitutes a major advance in predictive power over such earlier studies.

An important block of related evidence stems from the fact that my theory not only predicts that members of the RDP set will be deleted according to the derivations outlined in Tables 4.1 and 4.2, but makes the complementary prediction that predicates **not** belonging to that small set will of necessity remain as lexical items in the corresponding surface expressions. Corroboration of this prediction is to be found in an extensive and systematic set of expressions with precisely the predicted structure, namely, NPs containing compound adjectives of the form [N + participle] or [Prep + N] in prenominal position. These are, of course, exactly the structures produced by the derivations in Tables 4.1 and 4.2 when RDP Deletion rules are blocked from applying by the presence of a "nonmember" predicate. Examples of such expressions are given in (4.59).[34]

(4.59)   a. *communist-backed revolution*   *student-dominated groups*
             government-assisted projects   *battery-powered flashlight*
             university-initiated actions   *moss-covered rocks*

        b. *fish-eating dinosaurs*   *energy-wasting processes*
             party-loving women   *muscle-building exercises*
             meaning-bearing elements   *pennant-winning teams*

        c. *out-of-town performances*   *between-meal snacks*
             under-the-table deals   *across-the-board increases*
             over-the-counter sales   *off-campus housing*

In each case, the NP can be derived by the steps outlined earlier in Tables 4.1 and 4.2; however, since the predicates in these examples do not belong to the RDP set, no RDP Deletion rule can apply and the derivation is essentially completed at Step 8 of Table 4.1 (i.e., following Predicate Preposing).

Since the RDP Deletion rules apply in these derivations before Morphological Adjectivalization, the former cannot be sensitive to whether the latter will have affected the final, surface form; consequently, an

---

[34] I have no explanation for the fact that the set of compound adjectives illustrated in (4.59c) frequently requires a third morpheme (usually *of* or *the*) between the basic preposition and the noun.

additional prediction is made that the members of the RDP set will be deleted without regard to whether the surviving nominal element gets turned into an adjective or remains a noun. If this prediction is accurate, these derivations should produce comparably large sets of N–N CNs and Adj–N CNs whose underlying structures all share the same (deletable) predicate. The empirical support for this prediction has, in effect, been accumulating throughout this chapter, for the discussions of individual RDPs have in each case included examples of both N–N and Adj–N forms derived by deletion of the same predicate; the more extensive listings in the Appendix attest further to the derivational parallels. (Although by this point the near identity of the derivations for the two surface types may strike the reader as self-evident, it remains the case that few earlier analyses had explicitly linked these two sets of CNs and none had done so in terms of explicit rules situated within a systematic derivation. For this reason, the corroboration of this particular prediction is a valuable additional source of support for the proposal as a whole.)

Also relevant is the fact that the formulation of the RDP Deletion rules (in conjunction with others in the derivation) permits us not only to predict, but also to explain, the difference in predicability that is observed between the compound adjectives that are the input to RDP Deletion, and the nominal adjectives that are produced from the nouns in their output. The explanation lies in the constituent structure of the immediately antecedent subtree from which each adjective type is derived. The compound adjectives that are acted upon by RDP Deletion rules always consist of a verbal element and a nominal one, analyzable into one of these three structures:

(4.60)   a. passive participle + agentive (or instrumental) phrase
        (e.g., *caused by viruses* > *virus-caused*)
      b. active participle + object NP
        (e.g., *making music* > *music-making*)
      c. preposition + object NP
        (e.g., *in tówn* > *ín-town*)

Note, however, that the antecedents to these adjectives (i.e., the italicized noncompound phrases in [4.60]) are such that each comprises, together with a requisite copula, all the components necessary to make up a well-formed predicate phrase on the surface; and since the compound adjective consists of precisely the same lexical items (excluding the case marker *by*) as the antecedent periphrastic structure, it is to be expected that the compound adjective should also constitute (again, with a requisite copula) a well-formed predicate phrase for a surface sentence. It therefore should be grammatical in predicate position, and indeed it is.

In contrast, the antecedent to a nominal adjective formed within a CN is nothing more than the sole surviving member of the compound adjective, namely, its nominal element. When that adjective loses its predicate element (participle or preposition), however, it apparently also loses the predicating quality that the compound adjective as a whole possessed; what is left is a nonpredicator constituent (i.e., an N) which then is adjoined to the original head noun to become a single constituent. But since nominal adjectives are simply superficial variants of these non-predicating "survivor" nouns (and are, moreover, incorporated under a new N node), their nonpredicability follows directly from this derivational history. The RDP Deletion rules are further supported, then, by the fact that the structural change they effect provides an **explanation** for the significant difference in predicability observed from the start between compound adjectives and other "true" adjectives, on the one hand, and nominal adjectives in CNs on the other.

Still another source of corroboration for the RDP set and the RDP Deletion rules is related to the distinctive nature of the individual members of the RDP set. Although I based my original determination of its membership solely on an analysis of the semantic relations that seemed to best characterize the CNs in my "corpus," I subsequently found intriguing and potentially very important evidence that the members of this set have more in common than simply their deletability in CN formation. This evidence, which suggests that membership in the RDP set is limited to predicates of a universally "primitive" nature, is significant enough to merit discussion in a separate section (Section 4.5), since its implications take us far beyond the English-specific data reported on thus far, to the possibility of establishing **universal** semantic principles for the formation of CNs as well as for functionally equivalent structures (like the genitive) in other languages.

Since a more detailed discussion of this claim concerning universal grammar is provided in a subsequent section (6.4), let it suffice for the present to point out that important evidence from other areas of the grammar in which these nine predicates share distinctive syntactic and semantic characteristics gives us reason to believe that the membership in the RDP set may be independently motivated as well as universally significant.

The final argument that will be advanced here in support of the inclusion of an RDP Deletion process in these derivations is simply that such a solution appears to constitute the most satisfactory theory advanced to date for accounting for a large number of facts about both the semantic structures and the syntactic behavior of complex nominals. The proposals presented in these pages certainly go well beyond the traditional evaluation of CNs as "idiosyncratic" or variable without limit; more importantly, I claim that they are substantially more accurate than either earlier syntactically based analyses (such as Lees 1960 and Coates 1971) or previous analyses that included semantic facts as well (such as Marchand 1969 and Postal 1972).

On the other hand, no claim is made here that the present analysis is the only one possible for making the correct generalizations in this area; indeed, it is not difficult to imagine a lexicalist analysis that would in essence convert the syntactic derivations of Tables 4.1 and 4.2 into corresponding Semantic Interpretation Rules. (Such rules, rather than syntactically deleting the RDPs from derivationally antecedent forms, would presumably combine the meanings of these predicates in some manner with the meanings of the surface components of the CN in order to construct appropriate and complete interpretations for each surface form. See, for example, the discussion of noun compounds in Jackendoff 1975:656.) Whether such an approach would permit as satisfactory and unified an explanation for the numerous syntactic and semantic phenomena accounted for here in an integrated fashion is another question.[35]

4.4 Multiple Ambiguity Disambiguated

One important consequence of the theory presented here for deriving CNs by predicate deletion is that any CN so derived is potentially 12-ways ambiguous.[36] This figure follows from the fact that a CN is potentially

[35] This question cannot be answered at present, however, since no lexicalist has yet proposed in any detail just what an analysis of CNs based on semantic interpretation rules (SIRs) would comprise. Thus, Aronoff's *Word Formation in Generative Grammar* (1976) explicitly excludes compounds (its comprehensive title notwithstanding), while Jackendoff (1975) devotes just three pages to the topic. Although Jackendoff's discussion (1975:655–658) does include three sample "semantic redundancy rules," they are unfortunately of no greater sophistication than the corresponding transformational rules in Lees 1960 on which they are based. (For criticism of the latter, see Levi 1975, Section 6.1.2.) Moreover, Jackendoff confuses the notions of "familiar" and "possible" compounds by marking CNs such as *garbage tree* and *ant man* with asterisks (despite the obvious contemporary parallels of *junk sculpture* and *spider man* in the areas of art and children's comics, respectively), overestimates the significance of exocentric compounds like *redwing* and *yellow jacket* to the choice between lexicalist and transformational treatments of compounds, and makes the fallacious claim that the inclusion in the lexicon of separate entries for each "actually occurring compound" (in addition to more generalized morphological and semantic "redundancy rules") is an adequate—or even feasible—approach to the grammatical description of these forms. (For arguments against attempting to include each "actually occurring compound" in the lexicon, see Zimmer 1964:32 as well as Sections 3.2 and 6.2 of this book.)

[36] A more precise statement would have to specify several additional degrees of am-

9-ways **semantically** ambiguous since any one of the 9 RDPs may have been deleted; in addition, for the 3 predicates CAUSE, HAVE, and MAKE, the CNs are 2-ways **syntactically** ambiguous since the prenominal modifier could be derived from either the subject or the object of the underlying predicate. The 12 potential readings are in fact represented by the 12 groups of CNs appearing earlier in (4.1).

In actual usage, however, this relatively large ambiguity is reduced to very manageable proportions by judicious exploitation of semantic clues, lexicalization, and pragmatic or extralinguistic (encyclopaedic) knowledge. Since a more complete account of these strategies might well occupy several psycholinguistic monographs, the discussion here will be limited to illustrating each of these three approaches with a few concrete examples.

An example of semantic disambiguation is the presumption that the predicate IN can be the deleted predicate only if (*a*) the prenominal modifier can be construed as a locative noun (in either a concrete or abstract sense), and (*b*) the resultant locative relation between the head noun and its modifier seems semantically plausible in a given context. Thus, *marginal note, sea turtle,* and *marital sex* can all be interpreted easily on a locative reading since *margin* and *sea* are normally understood as concrete locative nouns, and *marriage* (> *marital*) can function as an abstract locative. On the other hand, it is as unlikely that we would assign locative interpretations to CNs such as *malarial mosquitoes, musical clock,* and *cough syrup* as it is that we would use their overtly locative paraphrases *mosquitoes in malaria, clock in music,* and *syrup in coughs;* for just the same reason that the latter expressions would be judged at least unusual, if not downright bizarre, in most contexts, listeners would also be reluctant to assign a locative interpretation to the corresponding CN "descendants."

Note, however, that a locative reading for *musical clock* would be ruled out because of the anomaly of the CN as a whole on such a reading, rather than because *music* could not be construed as a locative noun; in other CNs, such as *musical interval* or *musical innovation*, the noun *music* can appropriately be read as an abstract locative noun since the underlying structures of *interval in music* and *innovation in music* are, semantically, perfectly plausible.

Lexicalization also plays a part in helping us determine (or, more

---

biguity, since the possibility of deriving a prenominal modifier from a subject or an object of a nominalized verb adds at least two more readings to the ambiguity; moreover, CNs cannot be distinguished automatically on the surface between those derived by predicate deletion and those derived by predicate nominalization. What is important here, however, is that the principles of disambiguation remain the same.

accurately, make an educated guess at) just which reading of an unfamil-iar CN is intended by our interlocutor. This process is indeed indispensa-ble in understanding the semantic relation that underlies CNs like *water bed*, *pipe dream*, *head lettuce*, *lifeboat*, or *battle fatigue*; without knowing the specific, lexicalized senses of such forms, we would be likely to try to interpret them along the lines of comparable CNs like those shown in (4.61), and in so doing, misinterpret the intention of the person who used them in their lexicalized sense.

(4.61)

Lexicalized CN	Its RDP	Comparable CN	Its RDP
water bed	HAVE$_1$	*camp bed*	IN/FOR
pipe dream	CAUSE$_2$	*examination dream*	ABOUT
head lettuce	BE	*brain food*	FOR
lifeboat	FOR [*saving*]	*death trap*	CAUSE$_1$
battle fatigue	CAUSE$_2$	*morning fatigue*	IN

For example, since people normally think of beds as inherently dry objects, a foreign (but English-speaking) visitor who had neither seen nor heard of the object we now call *water bed* would have great difficulty in understanding just what this strange CN might refer to; certainly *bed in water*, *bed making water*, *bed for water*, or almost any of the other possible interpretations would strike her as inherently implausible. She would then have to be told "what a water bed was," that is, what its lexicalized meaning was. Thenceforth, she would associate this CN with a quite specific definition along the lines of 'bed with water-filled mattress' rather than with the more basic definition of 'bed which has water [as an essential part]', which the present theory could predict. To a limited extent, then, learning lexicalized readings of particular CN forms is one strategy by which we avoid having to make a number of potentially inaccurate guesses as to the semantic relation underlying an unfamiliar CN; although this is particularly valuable in cases where semantic clues are uninformative or pragmatic clues unavailable, it is an inherently limited strategy since it works only for culturally shared and noncreative readings of CNs.

The third basic strategy is that of pragmatic disambiguation, in which we apply extralinguistic (sometimes called "encyclopaedic") knowledge to identify the most plausible reading of a certain CN; such knowledge may be tied specifically to the discourse subject, to our knowledge of the personal background of the speaker using the CN, or, more generally, to our knowledge of our culture. For example, our normal interpretation of *musical clock* as 'clock that makes music' is presumably due to our experience with a variety of mechanical devices (such as tops and other toys, or clock–radios) which can produce music but which are in fact neither activated nor powered by music (i.e., they do not USE music to function in the way that *electrical clocks* must USE electricity to function). I claim, however, that this is a reflection of our present technology rather than our linguistic competence, and that **if** a device, such as a clock, could be made to function using music as its energy source, we could then properly call this device a *musical clock*, generating this new reading in exactly the same way as we now derive *electrical clock* or *steam iron* (i.e., by deleting the predicate USE).

A somewhat different example of the role of extralinguistic information in disambiguating CNs is afforded by the appearance in a recent financial column (in the *Chicago Sun-Times*) of the CN *vehicular vermin*. Out of context, this CN might be disambiguated by using such clues as the relative sizes of vehicles and vermin, and the common CN pattern that names living creatures after their habitats; one might thus assign a locative reading of 'vermin in vehicles' to this CN, along the lines of the comparable forms *barn vermin* or even *terrestrial vermin*. In proper context, however, a locative reading based on IN Deletion would have been anomalous, given the fact that the subject of the column was a proposed automobile plant to be built in the United States, and the full sentence in which the CN appeared was this one: "If VW decides to gestate its vehicular vermin here, they could come from anywhere." This use of *vehicular vermin* as a CN derived by BE Deletion (i.e., from the same source as *vehicles which are vermin*) is a creative one whose successful interpretation depends not only on the context provided by the printed text, but also on the reader's extralinguistic knowledge that VW cars are often referred to in this country as *bugs*. Without this extralinguistic knowledge, it is unlikely that readers of the column would be able to associate with this CN the same interpretation that the columnist presumably intended.

In summary, we use both semantic and pragmatic information, along with varying degrees of lexicalization, in trying to decide on the particular reading of a CN that an interlocutor is **most likely** to have had in mind. (Of course, a first hypothesis may have to be revised if that interpretation fails to fit coherently with the rest of the discourse, but this is true of any potentially ambiguous construction we may hear.) All of these (potentially overlapping) strategies contribute to the functional success of CN forms in natural languages by reducing to an essentially insignificant degree the multiple ambiguities that are inherent in such a concise and serviceable target structure.

4.5 Membership Requirements for the RDP Set

The thrust of the present chapter is contained in the claim that RDP Deletion rules must be written so as to delete only those predicates belonging to this very small set: CAUSE, HAVE, MAKE, USE, BE, IN, FOR, FROM, and ABOUT. We must now consider the very basic question raised by this claim: Why just these predicates? What quality is it that entitles these predicates, and no others, to membership in such an unusual and severely restricted set?

In attempting to construct a reasonable answer, we must observe to begin with that these predicates seem to embody some of the most rock-bottom-basic semantic relationships expressible in human language; assuming that there **are** such things as semantic primes, these nine (or at least an impressive majority of the nine) are surely outstanding candidates. In addition to this semantic distinctiveness, many of the RDPs manifest a high degree of syntactic distinctiveness as well, in the sense of belonging to comparably "exclusive" but syntactically defined special groups. The evidence for the syntactic distinctiveness of these RDPs comes from two different areas of the grammar, involving what may be called the features of "surface invisibility" and "grammatizability."

"Surface invisibility" characterizes those predicates that can be either deleted or incorporated before they reach the surface **without leaving a surface trace.** This term of "surface invisibility" is introduced here as a cover term that can (*a*) include or subsume the two processes of deletion and incorporation as defined in Gruber 1965 (especially pp. 13–17), and (*b*) still preserve the essential, significant difference between them. I know of no other term in the literature that would serve both these purposes.

It turns out that between four and six of the nine predicates in question have this ability to become invisible by Surface Structure. In normal English sentences, BE, IN, and HAVE are deletable, and CAUSE is incorporable. In addition, the predicate FOR appears to be deletable in the very special and restricted linguistic domain of signs, as in (*FOR*) *EMPLOYEES ONLY* on a door, or (*FOR*) *GREEN GLASS* above a recycling bin; in a somewhat similar manner, we might interpret titles appearing on physical objects (e.g., *Pioneer Women* or *Gothic Cathedrals* on a book) as instances of deletion of the predicate ABOUT (as well as of certain other predictable elements). Examples of each of these "disappearing acts" follow.

To begin with, the predicate BE is regularly deletable in English not only through WH-*be* Deletion (Relative Clause Reduction), but also in a variety of other constructions involving complements, apposition, and comparisons, as shown in (4.62).

(4.62) a. *The committee considers Ms. Abzug (to be) a most suitable candidate.*
  b. *Certain fools prefer their wives, like their dogs, (to be) subservient.*
  c. *They certainly* **seem** *(to be) well adjusted, don't you think?*
  d. *Whatever your opinion of him (is), you must admit he's tenacious.*
  e. *We were rated on teaching, research, and "good citizenship"—the classic three duties.*
  f. *She loved to introduce "her granddaughter the professor" to all her friends and acquaintances.*
  g. *Adar is as talented as Ariella (is).*

The locative predicate IN is deletable in English in a great many time expressions and in a few place expressions, as (4.63) indicates.[37]

(4.63) a. *The only time she's ever seen him was* $\begin{Bmatrix} \textit{(*on) last night} \\ \textit{(on) Tuesday} \\ \textit{(*in) last spring} \\ \textit{(on) that Saturday} \end{Bmatrix}$.

  b. *I'm (*at) home, dear.*
  I'm usually (at) home (in the) mornings.
  *They both work (*in) downtown.*

Although the predicate HAVE seems to be less generally deletable than either BE or IN, the sentences in (4.64) suggest that there is at least one environment in which the lexical item *have* is regularly deletable, namely, in the complement of verbs such as *want, require,* and *desire;* numerous arguments for this claim may be found in McCawley 1974 and in Ross 1974b.

(4.64) a. *Max wants (to have) a lollipop.*
  b. *Sue wanted (to have) your apartment until next summer.*
  c. *Joe wants (to have) some horses, but his mother won't allow it/*them.*

In addition to the three predicates BE, IN, and HAVE, the predicate CAUSE can also become invisible on the surface, but here it is a clear case of incorporation. It is well known that the semantic element CAUSE leaves no surface marker on many causative verbs in English, leading to

---

[37] The relative rarity of "invisible IN" in place expressions suggests that the few examples cited may be better analyzed as representative of idiosyncratic incorporation (again in the sense of Gruber 1965) rather than of a more general process of deletion. Whatever the solution in these cases, however, the more general quality of "invisibility" still obtains.

homonymy (though not true ambiguity, since the number of arguments
is distinctive for each) with related inchoatives, as may be seen in (4.65).
CAUSE is apparently also incorporated into certain adjectives of human
emotion, such as *sad* or *cheerful* in (4.66), which can be ambiguous
between an experiential reading and a causative reading.

(4.65) a. *He opened the door.*   versus   *The door opened.*
        b. *She reddened her cheeks.* versus   *Her cheeks reddened.*
        c. *They melted the snow.*   versus   *The snow melted.*

(4.66) a. *That sure is a sad/cheerful story!* [causative]
        b. *That sure is a sad/cheerful child!* [experiential]
        c. *That sure is a sad/cheerful character!* [ambiguous]

In addition to these instances of deletion and incorporation in normal
English sentences, it seems that a regular deletion process takes place in
the special context of signs. Many signs (though by no means all) can be
systematically analyzed as having the meaning of 'This x is FOR y', where x
is the object or place on which the sign is hung, and y is the word or words
actually visible. Examples of signs that can be so interpreted are given in
(4.67), together with sample contexts.

(4.67) a. WOMEN   (= 'This bathroom is for
                                   women.')
        b. TICKETS   (= 'This counter is for tickets.')
        c. ADMISSIONS   (= 'This office is for admissions.')
        d. MEN'S SHOES   (= 'This store is for men's
                                   shoes.')
        e. WORK CLOTHES   (= 'This laundry chute is for
                                   work clothes.')
        f. SUGGESTIONS   (= 'This box is for suggestions.')
        g. EMPLOYEES ONLY   (= 'This entrance is for em-
                                   ployees only.')

What is intriguing about these data is that they demonstrate a con-
tinuity both with analyses of predicate deletion in normal linguistic contexts (as analyzed throughout this work, for example) and with analyses of
deletion in other special contexts (such as the "label language" reported
on in Sadock 1974). Specifically, these data must be given an analysis
which includes the same feature of inherent **vagueness** that was shown
earlier to characterize CNs derived by FOR deletion; in both cases, although a particular verb could be specified for each surface form (e.g.,
the sign WORK CLOTHES might be derived from the same source as
*This chute is for **depositing** work clothes*), such specification would need-

lessly and improperly restrict the possible interpretations of the surface expression. Just as *bull ring* could mean 'ring for fighting bulls', 'ring for holding bulls', or 'ring for washing bulls' (as argued earlier), so the sign *WORK CLOTHES* could be used appropriately to mean 'This chute is for **depositing** work clothes' (on a kibbutz laundry chute, for example) or 'This store is for **selling** work clothes' (in a store window) or even 'This area is for **making** work clothes' (in a garment factory). What is common to all these possibilities is that the specific verb, if any, is predictable from the object on which the sign appears and so need not be incorporated specifically in a **linguistic** description; on the other hand, the deletion of the predicate element FOR is a regular, recurring phenomenon and so does belong in a linguistic description of this particular kind of "sign language."

The data in (4.67) are also consistent with a syntactic regularity noted by Sadock (1974) for linguistic usage on labels, namely, that the N or NP that refers to the real-world object to which the label or sign is attached may be regularly deleted. Consequently, to convey the message that 'The counter to which this sign is attached is for (buying) tickets', we need only write the nominal object of the predicate FOR on that sign (i.e., *tickets*); FOR is systematically deleted, and the rest will be deduced by pragmatic considerations.

In a still more restricted context, namely, that of printed titles, the predicate ABOUT is regularly deleted. Thus, while titles of books and other printed materials sometimes say what they **are** (e.g., *A Grammar of Old Norse, A Topographical Atlas of Sri Lanka*), in other cases they clearly designate not what those objects ARE, but what they are ABOUT; examples of this sort include titles like *Gothic Cathedrals, Power Volleyball*, or *The Role of Chinese Cooking in Linguistic Graduate Programs*. Just as the FOR signs discussed earlier make no mention of the objects on which they appear, so here the titles do not say "This book is about x" but only "x." (Note that the predicate ABOUT is not always deleted, since we also find titles, especially in technical or academic publications, like these: *On Human Aggression* or *Concerning a New Social Order.*)

Returning to the more central issue of "invisibility" as it relates to the RDP set, I have shown that four of the nine predicates that constitute this very exclusive set (as well as two more, within their own special contexts) share membership in another highly restricted set, namely, the set of predicates which need not have overt representation in Surface Structure. It seems unlikely that this is coincidental.

The second source of syntactic evidence relevant here is the fact that a number of these predicates tend to become "grammatized" in natural languages; that is, they are often expressed not by independent lexical

4 CN DERIVATIONS BY PREDICATE DELETION

items but by bound grammatical morphemes. For example, CAUSE is frequently expressed not by an independent lexical item, but by a grammatical marker on the verb to indicate that the semantic element CAUSE has been incorporated into the basic (or inchoative) verbal stem. Thus, Japanese uses the infix *-(s)ase-* in productive causative constructions, while Semitic languages such as Hebrew and Arabic reserve distinctive conjugational patterns (utilizing both affixation and vowel changes) to mark their productively causative verbs.

Other predicates in the RDP set are grammatized into markings on **nouns,** that is, into case endings. This is true of USE, IN, FOR, and FROM, which may surface as Instrumental, Locative, Dative or Benefactive, and Ablative (Source) markers on their respective objects. In addition, the predicate HAVE may become a Possessive case marker on either its subject, its object, or both. (For example, English and Latin mark the subject of HAVE, as in *a woman's life* or *vīta fēminae*; a few forms in Hebrew mark the object, as in *baal-a miryam* 'husband-her Miriam = Miriam's husband'; and Turkish marks both subject and object, as in *kadın-ın hayat-ı* 'women-POSS life-POSS = a woman's life' or *kız-ın kitab-ı* 'girl-POSS book-POSS = the girl's book'.) We might also consider the nominative case that marks certain appositive constructions (e.g., Latin *Martiālis poēta* 'Martial the poet') as a surface reflex of an underlying BE.[38]

It thus appears that the set of semantic predicates that can be expressed by bound grammatical morphemes is also made up of semantically very **basic** elements and so constitutes a third set with highly selective membership. Once again, it is unlikely to be a coincidence that the membership of these three sets, indicated by the check marks in (4.68), overlaps as much as it does.

(4.68)

	CAUSE	HAVE	MAKE	USE	BE	IN	FOR	FROM	ABOUT
RDPs	√	√	√	√	√	√	√	√	√
Invisible in SS	√	√			√	√	√		√
Grammatizable	√	√		√	√	√	√	√	

[38] Eric Hamp has pointed out to me (personal communication) that the members of the RDP set are strikingly parallel to the reconstructed case system of Proto-Indo-European. The correspondences he proposes are these: both CAUSE and FROM as ablative (suggesting that the two may be collapsible into a single relation), HAVE as possessive genitive, FOR as dative, USE as instrumental, IN as locative, MAKE as a sort of accusative, and BE as nominative. (ABOUT remains to be accounted for.) These correspondences, albeit not perfect, certainly lend weight to the hypothesis that the RDPs form a universally distinctive set, and suggest a promising line of investigation for subsequent studies of the universality of the RDP set.

The only conclusion that can be drawn at present from these data is that all three sets presented here seem to share at least one requirement for membership, namely, that hard-to-define but intuitively salient feature of "rock-bottom" semantic basicness. In addition, we may note that the exclusion of the indisputably basic predicates AND and NEG from the RDP set (in contrast to the inclusion of NEG in the grammatizable set, and of AND in both the others just discussed) means that this feature of semantic basicness must be considered to be a necessary, but not sufficient, condition for defining predicates as RDPs. Although we will return to the question of the universality of the RDP set in Chapter 6, the problem of determining the precise conditions that define membership in the RDP set has yet to be solved.

5

DERIVATIONS OF COMPLEX NOMINALS BY PREDICATE NOMINALIZATION

5.1 Types of Nominalizations

We turn now from an examination of CNs derived by predicate deletion to a consideration of CNs derived instead by predicate nominalization. This second major group is composed of CNs whose head noun is a nominalized verb, and whose prenominal modifier is derived from either the underlying subject or the underlying direct object of this verb (or, for "multi-modifier" CNs, from both of these NPs). We may therefore classify the relevant data first according to categories determined by the head noun, and second, according to categories determined by the prenominal modifier.

5.1.1 Classification of Nominalizations

Since the head noun in these CNs is always a nominalized verb, a classification of CNs according to head noun presupposes a coherent analysis of the kinds of nominalizations that regularly occur in English, whether or not these forms surface as component parts of CN structures. I propose that the four nominalization types named and illustrated in (5.1) (as well as in the Appendix) represent four fundamental distinctions

that must be drawn in any such analysis.[1] (Although my specification of
the four types is based on the meaning and source of the head noun
alone, full CN forms are provided to help clarify the intended reading of
each head noun. Note also that my use of the terms "Agent" and "Patient" for nominalization types is without regard to animacy of the intended referent.)

(5.1)   a. Act Nominalizations:
            *parental refusal, dream analysis, staff attempts, enemy
            invasion, birth control, musical criticism*
         b. Product Nominalizations:
            *musical critiques, oil imports, clerical errors, editorial
            comment, constitutional amendment, royal orders*
         c. Agent Nominalizations:
            *city planner, financial analyst, sound synthesizer, car
            thief, garbage compactor, film cutter*
         d. Patient Nominalizations:
            *student inventions, designer creations, mammalian
            secretions, senatorial nominees, presidential appointees, city
            trainees*

Although I shall not propose specific source structures for these four
types of nominalizations until the following section, it is important even
at this stage to understand the semantic differences among these types,
particularly since they are often obscured by the limits of English morphology. The paraphrases provided in (5.2) are thus intended to characterize informally the basic meaning of each of the four types; in (a) and
(b), two examples are given in order to illustrate CNs whose prenominal
modifiers represent either the subject or the object of the nominalized
verb, whereas in (c) and (d) the examples are chosen to illustrate CNs
with either animate or inanimate referents.[2]

---

[1] I leave open the possibility that at least one more type of nominalization may be
justified, namely, a nominalization derived from an NP complement construction with a
head noun of /STATE/ rather than /ACT/ or /PRODUCT/. This may be an appropriate source for
nominalizations like those discussed in Postal 1974 (e.g., *continuation, persistence,* and
tendency, which appear as head nouns in CNs such as *institutional continuation, student
persistence,* and *parental tendencies*). Since, however, my own studies have not included
such forms (an omission more inadvertent than principled), I must leave their analysis to
another time. (Note that the particular nominalizations examined by Postal differ systematically from those analyzed in this chapter in that the former either permit or require
sentential complements [e.g., *parental tendencies to dominate their offspring, student
persistence in demanding tuition reductions*]. This may or may not prove relevant to a
determination of their appropriate derivational paths.) See also footnote 5.

[2] The reader is reminded here that the nouns which underlie the prenominal modifiers

(5.2)   a. Act Nominalizations:
    parental refusal   'act of parents refusing'
    musical criticism   'act of criticizing music'
  b. Product Nominalizations:
    human error   'that which is produced by
                               (the act of) humans erring'

    musical critique   'that which is produced by
                               (the act of) criticizing music'
  c. Agent Nominalizations:
    mail sorter   'x such that x sorts mail'
    film cutter   'x such that x cuts film'
  d. Patient Nominalizations:
    student invention   'y such that students invent y'
    presidential appointee 'y such that presidents appoint y'

While the Act and Agent nominalization types are probably familiar to most readers, the Product and Patient Nominalizations may be less so. Although the referents of the Product Nominalizations are difficult to characterize semantically (that is, with any precision that goes beyond the informal paraphrases indicated in [5.2b]), they seem to represent some·object—usually though not always tangible—that is produced as the result of a specific action or event. Thus, the Product Nominalization *error* in CNs like *human error* or *clerical error* refers to the **outcome** of an act of erring; this outcome may be tangible or visible (as in the case of arithmetic or typing errors) or intangible (as in the case of certain errors in judgment or, say, "errors of omission"), but in both cases the outcome is something that can be referred to as an "object" or entity distinct from the act that produced it.

Although Patient Nominalizations also refer in some sense to the "product" of a certain act or acts, they must be systematically distinguished from Product Nominalizations on fundamental semantic grounds, as may be seen by comparing the paraphrases provided in (5.2) for the two types. In cases where Patient Nominalizations are overtly marked as such by the suffix -*ee*, we may construct minimal pairs to

---

in all the CNs in this chapter are generic nouns, which presumably are unspecified for number until quite late in the derivation. (See McCawley 1976a for supporting argumentation.) As a result, my choice of singular or plural nouns in the CN paraphrases of (5.2) and throughout this chapter must be understood as dictated by surface syntactic requirements rather than by any semantic or logical necessity. (Note that surface CNs require the singular noun for almost all prenominal modifiers, whereas non-CN paraphrases tend to have modifiers in the plural, at least for count nouns; cf. *city planner* but *planner of cities*, *student judgments* but *students' judgments* and *judgments by students*.

highlight the difference between Patient and Product types. This is done in (5.3) and (5.4), where the Product Nominalizations in the (a) examples are clearly different in meaning from the Patient Nominalizations shown in the (b) examples. (Note also that *appointments* and *nominations*, although identical in form to the Act Nominalizations of their respective underlying verbs, cannot receive Act interpretations in these sentences.)

(5.3) a. *The managerial appointments infuriated the sales staff.*
       b. *The managerial appointees infuriated the sales staff.*

(5.4) a. *We expect to receive the senatorial nominations shortly.*
       b. *We expect to receive the senatorial nominees shortly.*

The difference between Product and Patient Nominalizations is also reflected by the possibility or impossibility of having a lexically specified direct object co-occur on the surface with the nominalized head noun. If we look at the CNs in (5.1b) and (5.2b) (as well as in Group N in the Appendix), we see that the meanings of Product Nominalizations such as *musical critique* or *constitutional amendment* are such that the underlying structure often contains an NP object with semantic content of its own; this object may emerge in surface structure either as a postnominal adjunct (*critiques of music, amendments to the constitution*) or as a prenominal modifier (as in the CNs cited). In contrast, the meaning of a Patient Nominalization such as *student inventions* ('things that students invent') or *presidential appointees* ('people that presidents appoint')[3] is such as to preclude the appearance in surface structure of any lexical item besides the head noun that would purport to represent the underlying object. That is, since the head noun itself not only conveys the meaning of the underlying verb but also denotes the referent of the underlying object NP, there is simply no way that another direct object could be present in either underlying or surface structure. To illustrate, we can predict the impossibility of such expressions as *student inventions of paper clips* or *presidential appointees of women* from the ill-formedness of the underlying structures from which they would have to be derived, namely, those structures that would also underlie *things that students invent paper clips* and *the ones that presidents appoint women*. The lack of coreferentiality between the heads of these relative clauses

---

[3] I have no clear idea as to how the distinction in animacy between *inventions* and *appointees* that is expressed in these paraphrases should be represented in underlying structure. Note, however, that the *-ee* suffix does seem to be restricted to animate patients; we thus have *draftee, evacuee, nominee,* and *trainee* but not **inventee, *secretee,* or **staplee* for things invented, secreted, or stapled. (Two exceptions brought to my attention by Georgia Green are *controllee* and *deletee*, found in recent syntactic literature; this extension of the *-ee* suffix to inanimate referents is, however, very rarely made.)

and the object NPs precludes their being transformed into Patient Nominalizations. In fact, it is just this requirement of coreferentiality between the head of the relative clause construction and its direct object NP that distinguishes the Patient Nominalization from the Product Nominalization. (As a consequence, the source structure for the former will be a relative clause containing a bound variable, whereas the source of the latter will be an NP complement structure, as will be shown in the following section.)

The distinction between Product and Patient Nominalizations is as important as that between Act and Agent Nominalizations. The fact that the former distinction has apparently escaped notice while the latter is universally recognized in the literature on nominalizations may be due, at least in part, to the obscuring influence of English morphology. Thus, although there is ample syntactic and semantic evidence to justify distinguishing the four basic types of nominalizations identified in the preceding pages, this four-way contrast is, as far as I know, **never** fully reflected in English by overt morphological distinctions. Instead, English morphology tends to conflate two or even three of the different nominalization types into the same morphological shape; complicating the picture still further is the fact that the Patient type manifests such restricted productivity that the lexical gap is more the rule than the exception. These obscuring tendencies are exemplified by the sets of nominalizations shown in (5.5); verbs whose morphology permits three-way contrasts are shown in (a); those limited to two-way contrasts are shown in (b).

(5.5)

	Act	Product	Agent	Patient
a.	*criticism*	*critique*	*critic*	*—*
	importation	*import*	*importer*	*import*
	nomination	*nomination*	*?nominator*	*nominee*
	control	*control*	*controller*	*?controllee*
b.	*invention*	*invention*	*inventor*	*invention*
	gift, giving	*gift*	*giver*	*gift*
	synthesis	*synthesis*	*synthesizer*	*—*
	editing	*editing*	*editor*	*—*
	destruction	*destruction*	*destroyer*	*—*

The four-way distinction in nominalization types which I am proposing here has not, to my knowledge, yet been recognized in any of the generative literature with which I am familiar.[4] In these studies, almost all

[4] This literature includes Fraser 1970; Lakoff 1970; McCawley 1968 and 1975; Newmeyer

cited forms represent Act Nominalizations on which the discussion accordingly centers; in most cases, little or no recognition is given to the fact that these constitute only one significant type in the actual data. As for the other three types, Agent and Patient Nominalizations are discussed only by Lakoff (1970) and McCawley (1968, 1975) and then rather briefly, whereas Product Nominalizations are not discussed explicitly anywhere at all.

It is both a surprising fact and an ironic circumstance that, despite the crucial role that the analysis of nominalizations has played in the theoretical controversy between generative semantics and lexicalism (a controversy amply documented by the arguments and counterarguments to be found in the works cited in footnote 4), so much energy has been put into metatheoretical debate when so little detailed analysis of the data itself has been undertaken by either side. While the analysis offered in these pages is of course intended to remedy this situation at least in part, it too must be regarded as basically programmatic until more extensive supporting studies have been carried out.[5]

5.1.2 Classification of NOM CNs

CNs whose head nouns are derived by nominalization (henceforth, NOM CNs) may also be classified according to the syntactic source of their prenominal modifier(s). In this way the data may be divided into two major categories: Subjective NOM CNs, whose prenominal modifier derives from the underlying subject of the nominalized verb, and Objective NOM CNs, whose prenominal modifier derives instead from that verb's underlying direct object. In those cases where both subject and object emerge as prenominal modifiers within a CN, we have instances of a third composite type, which shall here be called the "Multi-Modifier" type. Examples of Subjective, Objective, and Multi-Modifier NOM CNs are provided in (5.6), (5.7), and (5.8), respectively (with additional exam-

---

1970, 1971, 1976; Postal 1972 and 1974; and Ross 1974a on the "transformationalist" side; and Chomsky 1970 as the major work on nominalizations on the "lexicalist" side. (In addition, the interested reader is referred to two recently published works which appeared too late for consideration here, namely, the lexicalist studies of Jackendoff 1977 and Selkirk 1977.)

[5] The reader may have noticed, for example, that the present work does not include any analysis of nominalizations of predicate adjectives or of the CNs in which they appear (e.g., *racial equality*, *popular indifference*, *student eagerness*). This exclusion is dictated by simple prudence, rather than by ideological conviction, since I have not investigated these forms in any separate detail. Nonetheless, it does appear plausible that such nominalizations could be derived from NP complement structures with a head noun such as /STATE/ or perhaps /QUALITY/, while the CNs in which they appear could be subsumed under the Subjective Nominalization category illustrated in (5.6). I shall, however, not pursue this topic further in these pages.

ples to be found in the Appendix); in each case, they are further divided to show the kind of nominalization that the head noun represents. (The reader is reminded, however, that these head nouns may be ambiguous among two or even three nominalization types, along the lines suggested by the chart in [5.5].)

(5.6)

SUBJECTIVE NOM CNs

Act: *parental refusal*
manager attempts
cell decomposition
judicial betrayal

Product: *clerical error*
peer judgments
faculty decisions
papal appeals

Agent: —

Patient: *royal gifts*
student inventions
presidential appointees
city employees

(5.7)

OBJECTIVE NOM CNs

Act: *birth control*
heart massage
subject deletion
musical criticism

Product: *oceanic studies*
chromatic analyses
stage designs
tuition subsidies

Agent: *draft dodger*
urban planner
acoustic amplifier
electrical conductor

Patient: —

(5.8)

MULTI-MODIFIER NOM CNs

Act: *industrial water*
pollution
parental child abuse
city trash collection
infant heat regulation
administration press
censorship

Product: *government price supports*
student monetary demands
municipal oil imports
executive stock purchases
college minority admissions

Agent: —

Patient: —

The gaps that appear in (5.6)–(5.8) are wholly systematic ones. It was pointed out earlier that the meaning of a Patient Nominalization precludes its co-occurrence with any independent NP modifier representing an underlying object; as a result, the set of Objective NOM CNs shown in

(5.7), whose prenominal modifiers are always derived from underlying object NPs, can never have a Patient Nominalization as head noun. The absence of Agent Nominalizations among the Subjective NOM CNs in (5.6) may be explained along precisely parallel lines since the underlying subject NP can be represented either by an Agent Nominalization or by a prenominal modifier like those found in the CNs of (5.6), but not by both.[6] Finally, since Multi-Modifier NOM CNs include prenominal modifiers derived from both the underlying subject and the underlying object, neither Agent nor Patient Nominalizations may be used for their head nouns, as indicated by the gaps in (5.8).

The chart in (5.9) provides a convenient summary of the distribution of all types of NOM CNs. Here we see that although the combination of four basic types of nominalizations with the three types of CNs determined by prenominal modification gives us 12 hypothetical categories, considerations of semantic and syntactic well-formedness reduce the actually occurring categories to 8.

(5.9)

	Subjective	Objective	Multi-Modifier
Act	*parental refusal*	*dream analysis*	*city land acquisition*
Product	*clerical errors*	*musical critique*	*student course ratings*
Agent	—	*city planner*	—
Patient	*student inventions*	—	—

5.2 Nominalization: The Crucial Rules

Full derivations for CNs derived by predicate nominalizations will be provided, rule by rule, in Section 5.3. In this section, however, I wish to isolate four of those rules—the nominalization rules themselves—so as to permit a lengthier discussion of some of the issues associated with their proper formulation.

The transformationalist analysis that I am proposing in these pages permits two different kinds of intermediate structures to serve as sources

---

[6] Although it is possible to find well-formed CNs like *student planners* and *women instructors* which **look** as if they might be "Subjective Agent" NOM CNs, it can be shown that these must be derived by BE Deletion (from structures corresponding to *planners who are students* and *instructors who are women*, respectively) and not by predicate nominalization. An analysis in terms of nominalization for forms like these would have the unacceptable requirement of positing source structures roughly of the form *ones who plan by students* and *ones who instruct by women*, respectively, which are clearly ill-formed on both semantic and syntactic grounds.

for the four nominalization types presented in Section 5.1.1. These source structures are (*a*) an NP complement construction as source for Act and Product Nominalizations; and (*b*) a relative clause construction containing a bound variable as source for Agent and Patient Nominalizations.[7] Although the input trees thus vary in important ways, the four rules of nominalization transform these structures into roughly isomorphic output trees (or subtrees, to be precise). The configuration of the output structures is shown in generalized form in (5.10). This shared subtree always consists of a highest NP (here, NP_4) whose leftmost NP daughter dominates the head noun newly formed by nominalization; that head noun in turn consists of the underlying verb adjoined as a right sister to a single morpheme such as /ACT/ or /AGENT/, shown in (5.10) simply as N_1. In addition, any other NPs which appear as daughters of NP_4 correspond to one or more of the arguments in the underlying S of the antecedent structure.

(5.10)

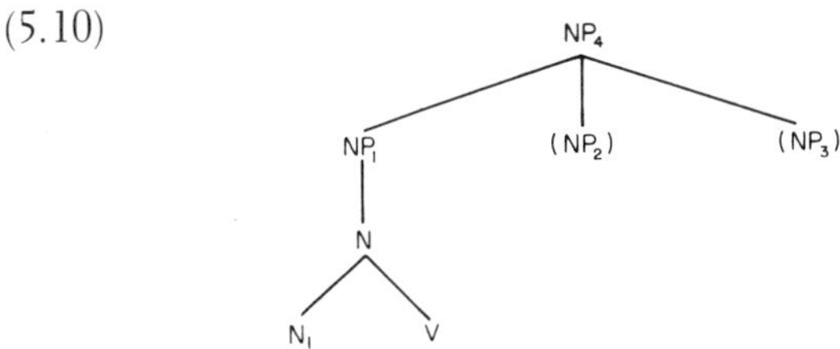

The four specific rules of nominalization that I am proposing here are illustrated in full detail in (5.11)–(5.14). From these diagrams the reader can readily observe not only the isomorphism of the output trees but also the structural parallels between the input trees for Act and Product Nominalizations on the one hand (in [5.11] and [5.12]), and those for Agent and Patient Nominalizations on the other (in [5.13] and [5.14]). In all cases, the format and notation anticipate subsequent usage in Section 5.3; specifically, NP_1 is used for the NP head of either the relative clause or the NP complement construction, NP_2 for the NP that dominates the case-marked subject NP, NP_3 for the NP dominating the case-marked direct object, if any, and NP_4 for the highest node in the entire source

[7] Although the transformationalist literature on nominalizations provides almost no detailed discussion of the actual rules of nominalizations, one can nonetheless find precedents for each of these proposed source structures in those few works that provide at least one relevant tree for public scrutiny. Specifically, an NP complement structure is proposed by both Fraser (1970) and Newmeyer (1970, 1976) as the source for Act Nominalizations, whereas a relative clause structure containing a bound variable has been suggested by McCawley (1968, 1975) as the appropriate source for **all** nominalizations (although McCawley's discussion includes only those three types that are here called Act, Agent, and Patient).

constituent. (Where additional clarification is needed, N_s, N_o, and N_v are used to mark nouns derived from the underlying subject, direct object, and verb, respectively.) Because only subject and direct object NPs are relevant to CN formation, no other NPs are shown in the tree diagrams of this chapter, and the terms *object NP* and *direct object NP* are used interchangeably.

(5.11) Act Nominalization

    Example: /ACT/ *## pollute by industry of water ## $\Rightarrow$*
          pollution by industry of water

Input tree:

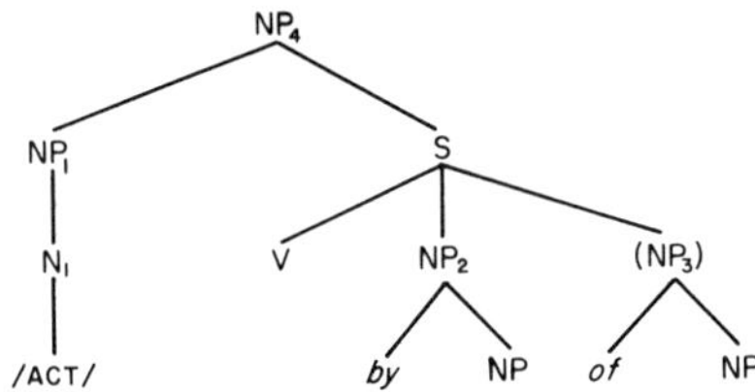

Output tree:

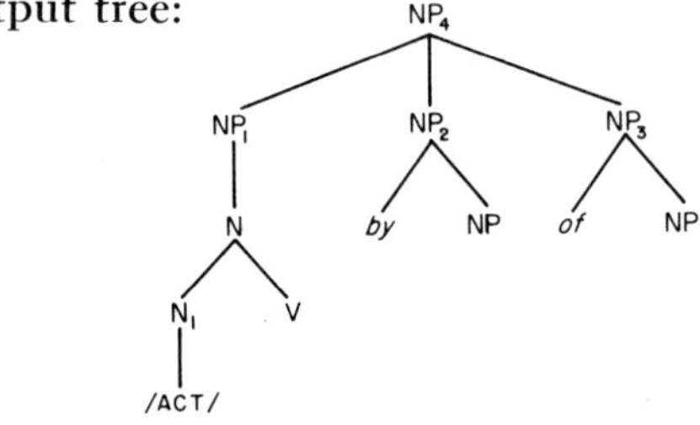

Comments

1. This rule applies optionally.
2. /ACT/ is a morpheme whose phonological shape will be spelled out by a later rule. If this rule of nominalization is not applied, /ACT/ must appear as the independent word *act*; if this rule is applied, /ACT/ will appear as a bound morpheme within the noun formed by nominalization (e.g., as *-tion* in *evaluation* or *-ment* in *development*).
3. The S node of the input tree is presumably removed by a structure-deleting convention which removes Ss that have come to dominate only NPs.[8]

---

[8] I take no position here on whether the loss of the S node is to be achieved by a tree-pruning convention along the lines suggested by Ross (1969:296, 299) or by well-formedness constraints to be imposed on intermediate trees along the lines suggested by Anderson (1976). What does seem essential, however, is to have the structure of these

(5.12)   Product Nominalization
         Example: /PRODUCT/ ## *criticize by students of music* ## ⇒
                critique by students of music

Input tree:

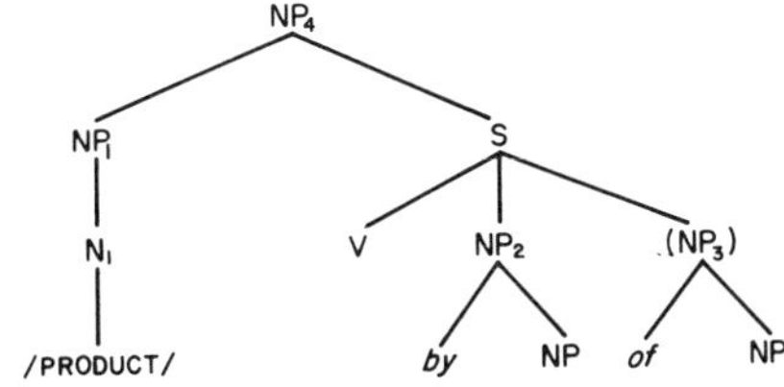

Output tree:

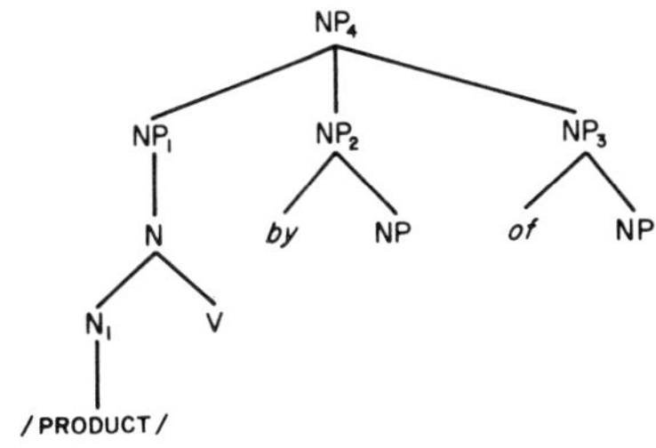

Comments

1. This rule applies optionally.[9]
2. /PRODUCT/ represents a morpheme whose phonological shape is spelled out by a later rule; in almost all cases in English, however, the noun formed by incorporating a verb with /PRODUCT/ is isomorphic with the noun formed by incorporating that verb with /ACT/.
3. The S node of the input tree is presumably removed by a structure-deleting convention which removes Ss that have come to dominate only NPs. (See footnote 8.)

(5.13)   Agent Nominalization
         Example: *x ## plan by x of cities* ## ⇒
                planner of cities

---

intermediate trees somehow reflect the fact that it makes no sense to say that the one or more NP nodes "left behind" by nominalization formation constitute a constituent of the type S (if indeed they constitute a single constituent at all).

[9] Actually, I have specified optional application only for the sake of consistency; it is, however, very difficult for me to imagine what surface structure would result if this rule were not applied (hence, the awkwardness of the paraphrases offered earlier in [5.2b], q.v.).

Input tree:

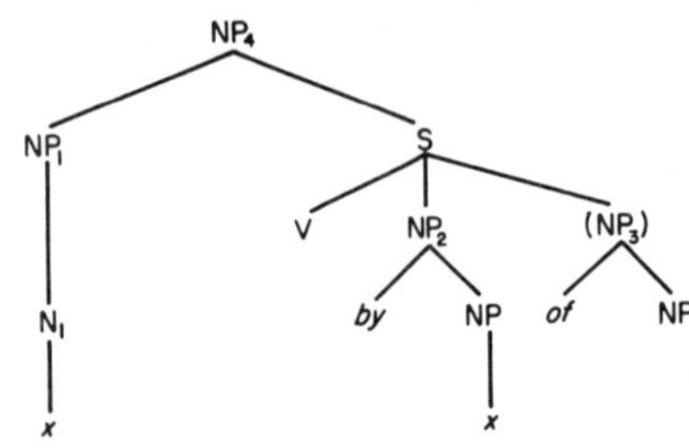

Output tree:

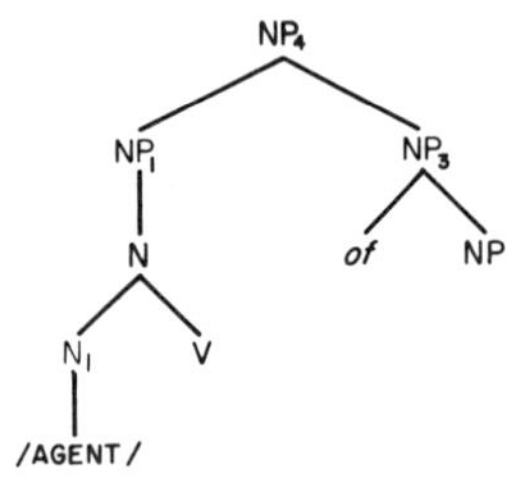

Comments

1. This rule applies optionally.
2. /AGENT/ is a morpheme whose phonological shape is spelled out by a later rule. In English it is almost always realized as a suffix on a verbal stem; its forms include *-er, -or, -yst,* and *-∅,* as in *manager, dictator, analyst,* and *cook.*
3. The S node that appears in the input tree is presumably removed by a structure-deleting convention which deletes Ss that have come to dominate only NPs. (See footnote 8.)

(5.14)   Patient Nominalization

Example: *y ## invent by students of y ## ⇒ inventions by students*

Input tree:

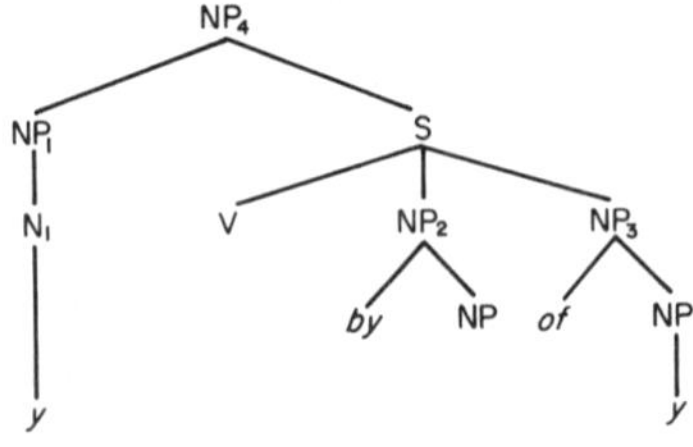

            5 CN DERIVATIONS BY PREDICATE NOMINALIZATION

Output tree:

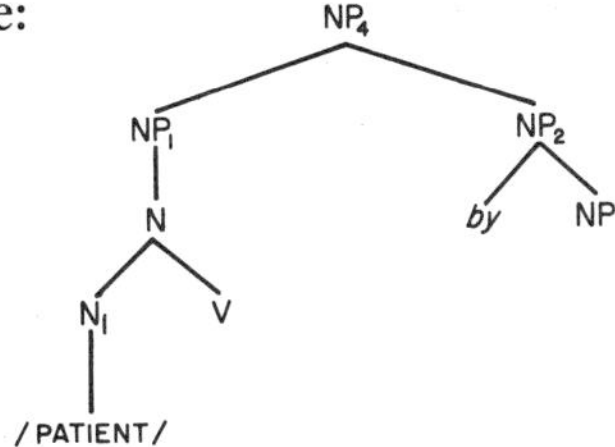

Comments

1. This rule applies optionally.
2. /PATIENT/ is a morpheme whose phonological shape is spelled out by a later rule. (In English, however, it is the exception rather than the rule for there to be a single lexical item for any Patient Nominalization, whether animate or inanimate. See [5.5] and Group P in the Appendix for examples.)
3. The S node that appears in the input tree is presumably removed by a structure-deleting convention which deletes Ss that have come to dominate only NPs. (See footnote 8.)

The reader may well be wondering at this point why a two-pronged approach has been adopted here instead of one in which all nominalizations derive from a closely comparable input structure. The simplest answer is that I do not find either relative clauses alone or NP complements alone to be satisfactory sources for all four of the nominalization types with which we are here concerned.[10] Moreover, the data themselves seem to divide into two related but distinct groups, with Act and Product Nominalizations forming one such group and Agent and Patient Nominalizations forming another.

The Act and Product types are similar in that (*a*) their meaning seems to include some semantic material that cannot be represented simply by a bound variable and an embedded S (i.e., seems to require a head noun that has semantic content—hence the /ACT/ and /PRODUCT/ heads for these two types);[11] and (*b*) they permit full lexical specification in SS of

---

[10] The fact that each of the analyses of nominalizations cited in footnote 7 was based on a single source type is, I believe, due in part to the highly programmatic nature of those studies, as well as to the fact that none of them attempts to account for all four of the nominalization types that have been documented here.

[11] Fraser (1970) and Newmeyer (1970) provide specific arguments for the choice of /ACT/ as the underlying head noun for Act Nominalizations, based on the similarities that may be observed between this nominalization type and surface phrases like *the act of* V-*ing*. What I

both subject and object NPs in addition to the head noun formed by nominalization. In contrast, the Agent and Patient types resemble each other (and differ from both Act and Product types) in that (*a*) the **only** satisfactory paraphrases available in English take the form of relative clauses with relatively unspecified heads (i.e., heads specified for no other feature but animacy and perhaps number); and (*b*) they each preclude the lexical specification of one particular underlying argument (i.e., Agent Nominalizations preclude the appearance of an independent subject NP, and Patient Nominalizations that of an independent object NP, as discussed earlier). We may note in addition that the meaning of Agent and Patient Nominalizations is such that in any semantic representation for these forms, there must be an argument in the embedded S that is coreferential to the head of the whole construction; in contrast, no such requirement of coreferentiality is forced upon us for the Product Nominalization and perhaps not for the Act Nominalization either.[12] These similarities between the Act and Product types on the one hand, and the Agent and Patient types on the other, make a two-pronged analysis like that suggested here a more plausible approach than one might at first believe.

This is not to say, however, that I am wholly satisfied with the asymmetry of my two-source analysis (or that I expect all my readers to be). In asserting that our present knowledge seems too limited to permit the elaboration of a uniform derivation for all nominalizations, I do not at all preclude the possibility that more extensive research into the subject may eventually justify a single source structure for all four types in question; indeed, I believe that the relative clause source proposed by McCawley is a much stronger candidate for this uniform source than the NP comple-

---

am proposing here is that a comparable approach may be used to justify the derivation of Product Nominalizations from a source with /PRODUCT/ as the head noun, despite the somewhat greater awkwardness of surface expressions such as *the product of NP's V-ing*.

[12] I must point out here that in an analysis such as that proposed by McCawley (1968, 1975), who suggests that "all nominalizations have semantic representations of the form 'the (an) x such that f(x)' [McCawley 1968:75]," there **would** necessarily be a requirement of coreferentiality for all nominalization types since all would in fact have relative clause constructions as sources. I see at least two difficulties with this proposal, however. The first is that I can find no coherent relative clause structure that comes close to expressing the meaning of Product Nominalizations in particular; the second is that it is not yet clear, at least to me, what linguistic consequences follow from an analysis like McCawley's in which verbs are subscripted with the bound variable that is also the head of the relative clause. (See McCawley 1968:75 and 1975:219 for a few more details of this analysis, as well as Davidson 1967 for some semantic arguments in its support.) It is for both these reasons that I have not been able to extend McCawley's otherwise persuasive analysis to apply in a uniform way to all four of the nominalization types considered here.

          5 CN DERIVATIONS BY PREDICATE NOMINALIZATION

ment I have proposed here for half the data. (That is, it is quite conceivable that future research will show that the NP complement structure which appears as "Input Tree" in [5.11] and [5.12] must be derived from a relative clause construction as a deeper source or, alternatively, that no nominalization derivation need include any intermediate structure having the configuration of an NP complement.)[13] These speculations notwithstanding, there does not seem to be adequate justification at the present time to impose a single-source analysis on the nominalization data presented here. Instead, I offer the two-source proposal suggested by the rules of (5.11)–(5.14) as the most adequate account of the data that I can produce, and hope that it will inspire other linguists either to refine it in the direction of greater accuracy and, if possible, greater generality, or to supplant it with a clearly superior analysis of their own.

Strictly speaking, of course, the task of providing an adequate account of nominalization processes in English lies outside the scope of this book, especially in view of the fact that the steps in CN formation that are essential here are those which **follow** the formation of the nominalized head noun rather than those that lead up to it; thus, this chapter is centrally concerned with the rules for transforming expressions such as *pollution by industry of water* into CNs such as *industrial water pollution*, and is much more tangentially concerned with the way in which the former expression is itself derived. I will therefore conclude this section with just a few additional remarks on the three nominalization types that have received **least** attention in the generative literature to date, namely, Product, Agent, and Patient.

As noted earlier, it is particularly difficult to find coherent paraphrases for the meanings of Product Nominalizations such as *municipal oil imports* or *executive stock purchase*. It is possible, however, that rather than opposing a Product Nominalization to an Act Nominalization (e.g., 'the "product" of executives purchasing stocks' as opposed to 'the act of executives purchasing stocks'), we need to recognize that the former **includes** the latter systematically; thus, a CN such as *executive stock purchase* would have the meaning 'the product of **the act of** executives

---

[13] Both of these approaches have been suggested in the literature for Action Nominalizations, at least. Thus, Ross's analysis (1972b) of the abstract verb DO suggests (implicitly) that the head noun /ACT/ might be analyzed as a Patient Nominalization of DO (that is, that /ACT/ may be derived from a structure such as this one: x: DO (y, x), where x is a sentential object of DO). The second alternative corresponds to McCawley's analysis of nominalizations (McCawley 1968, 1975) in which all nominalizations are claimed to be derived from relative clause structures. It is unclear, however, how either of these analyses might be extended to derive Product Nominalizations as well, and thus obviate the need for an NP complement source.

purchasing stocks'. In this way, Product Nominalizations would stand in relation to Act Nominalizations the way many causative verbs in English, at least, stand in relation to inchoatives. That is, just as many lexical verbs in English may express either an inchoative predicate or an inchoative plus a causative (e.g., *redden* can mean either 'become red' or 'cause to become red'), so it may be that a given surface nominalization may express either an Act Nominalization or a "Product-of-an-Act" Nominalization (e.g., *purchase* can mean either 'act of purchasing' or 'product of act of purchasing'). Whether this notion will be useful in finding a well-motivated source structure that more accurately reflects the meaning of the Product Nominalization is a question that cannot, unfortunately, be answered here.

As for the Agent and Patient types, the analysis represented in (5.13) and (5.14) follows the lines first suggested, as far as I know, by McCawley (1968:75). The use of a bound variable analysis like this one instead of one with an NP complement structure as source has the important advantage of providing a natural explanation of why one of the major arguments in the embedded S is always lexically empty, and hence why surface expressions such as *bank robbers by women* and *presidential appointees of crooks* do not occur. In contrast, an analysis of Agent and Patient Nominalizations that would derive them from NP complements (with /AGENT/ and /PATIENT/ morphemes already present as head nouns) could not prevent the derivation of such ill-formed expressions without imposing ad hoc restrictions on just which arguments could be lexicalized in the source structures for these two types and which ones could not.[14]

It is important to recognize that the set of NPs generated by the Agent Nominalization rule of (5.13), as well as the set of CNs that may be derived from those NPs, comprises not only NPs whose referents are animate beings performing volitional activities (i.e., referents subsumed under the more traditional use of the term "Agent") but also NPs whose referents are inanimate objects; thus, the rule of (5.13) will generate the head nouns not only for *urban planner* and *financial analyst*, but also for CNs such as *electrical conductor*, *photographic enlarger*, *acoustic amplifier*, and *metallic separator*.[15] The difference in animacy observable in

---

[14] The theoretical advantage just described for using bound variables to analyze nominalizations was first brought out by McCawley (1975:219) in a more general discussion of the value of a transformational treatment of nominalizations such as his in comparison with the lexicalist treatment suggested by Chomsky (1970).

[15] In noting that inanimate objects often serve as legitimate subjects of the abstract verb DO, Ross (1972b:105–106) has also emphasized the need to distinguish between the traditional category of **volitional** (animate) agents and the set of possible subjects of activity verbs. It is the latter, larger set which the Agent Nominalization rule of (5.13) is intended to generate.

     5 CN DERIVATIONS BY PREDICATE NOMINALIZATION

these CNs (cf. also *orchestra conductor* and *heat conductor*) is assumed
here to follow not from a difference in the transformations that produce
these forms but rather from other, relatively independent factors that are
not germane here. (These might include differences in the senses of the
underlying verb, a difference in head noun corresponding to that between *something* and *someone*, or a difference in speaker assumptions
about the customary referents of a certain Agent Nominalization; note
that some of these forms could have both animate and inanimate referents, as in *film cutter* or *mail sorter*.)

There is, however, one set of agentive forms involving CNs which the
rule of (5.13) is not intended to derive; these are the expressions (discussed earlier in Section 3.4) that consist of an agentive suffix attached to
a CN formed by regular processes; examples include *quantum mechanician, pastoral artist, symbolic logician,* and *criminal lawyer,* which I
claim are derived by partially idiosyncratic morphological rules from a
structure equivalent to *one who does quantum mechanics/pastoral art/
symbolic logic/criminal law.* (Some of the otherwise puzzling forms cited
in Brekle 1970 fall into this category as well: *drug experimenter, peace
marcher, manual laborer, fieldworker,* and probably *fan dancer.*) Since it is
the derivation of the CN **stem** in these forms that is my primary concern,
no explicit attention will be given here to the more peripheral rules
involved in deriving related nouns like *quantum mechanician* or, for that
matter, related adverbs like *quantum mechanically.*

In view of the other parallels between Agent and Patient Nominalizations, it is not too surprising that the Patient Nominalization rule of (5.14)
also serves to derive Patient nouns with both animate and inanimate
referents. In this case, however, the difference is reflected in surface
morphology; thus, when Patient Nominalizations are lexicalized as such,
only animate Patients receive the distinctively Patient-marking suffix *-ee*
while inanimate Patients generally appear as nominalizations in *-tion,*
-ment, or *-∅* (as in *invention, requirement,* and *request*) and thus are
ambiguous in form between at least an Act and a Patient reading. (See
footnote 3, however, for two exceptions.)

The interaction of semantics and morphology is, in fact, quite interesting in the case of Agent and Patient Nominalizations, although the
Patient type in particular deserves more thorough scrutiny than has been
possible here. One of the most puzzling aspects of the Patient noun is the
fact that although it is semantically the direct counterpart of the Agent
noun, the latter shows immense productivity in terms of direct lexicalization possibilities whereas the former shows a comparably immense **lack** of
productivity. On the one hand, Agent Nominalizations in *-er* or *-or* are
coined with great ease and frequency (thus increasing the very large
number of these forms already in circulation); in marked contrast, there is

almost no truly productive way of forming morphologically distinctive Patient Nominalizations with inanimate referents, and even the -*ee* suffix that forms Patient nouns for animate referents carries a certain stylistic taint of awkwardness or even crudity when extended beyond a relatively small number of common cases. (Consider, for example, the fact that there are no individual English words to name 'the thing read', 'the thing lost', or 'the thing stapled', although the comparable Agent nouns *reader*, *loser*, and *stapler* are in free circulation; in addition, although we do speak of *trainees* and *draftees*, the use of *hirees*, *promotees*, or *terminees* to describe those hired, promoted, or fired has simply not caught on.)[16] It remains to be seen whether any principled explanation can be found for the virtual exclusion from productive English morphology of a form that presumably has as straightforward a semantic structure as the Agent Nominalization with which it has been grouped here.

5.3 Derivations of NOM CNs

5.3.1 An Overview

Let us turn now to the details of the derivations by which we may derive the eight kinds of NOM CNs shown in (5.9). The five NOM CNs whose derivations are sketched in chart form in Tables 5.1–5.5 have been chosen to illustrate the operation of all relevant rules; thus, *parental refusal* exemplifies both Act Nominalizations and Subjective NOM CNs, *musical critique* both Product Nominalizations and Objective NOM CNs, *industrial water pollution* a Multi-Modifier NOM CN with a head noun based on /ACT/, and *urban planner* and *presidential appointee* the Agent and Patient Nominalization types, respectively.

The linear presentation of these tables is intended only to highlight the most obvious features of each transformation (such as major changes in constituent ordering and/or structure) so that the reader can get an overview of these derivations before considering each rule in more detail; thus, the constituent labels that appear as a matter of course in later trees have been omitted here, and the brackets used show only the most significant details of constituent structure and sisterhood relations. Complete information on the input and output structures of each rule is

---

[16] In a section concerned explicitly with Patient Nominalizations in -*ee*, Marchand (1969) points out that this group of nouns "has recently come into favor especially with words of official military jargon. Examples are *draftee* 1869, *enlistee*, *rejectee*, *selectee*, *trainee*, *evacuee*. . . . The suffix is a vogue morpheme which has formed many words of a more general application in present-day American English. But many of these have a playful nuance and a decidedly transitory character [p. 268]."

TABLE 5.1
Act Nominalization in a Subjective NOM CN

Transformation	Output structure
First cycle	/ACT/ ## REFUSE PARENTS *x* ##
1. Lexical Insertion	/ACT/ ## [*refuse*] [*parents*] [*x*] ##
2. Case Prep Spelling	/ACT/ ## [*refuse*] [*by parents*] [*of x*] ##
Second cycle	
3. Act Nominalization	[/ACT/ + *refuse*] [*by parents*] [*of x*]
4. Lexical Insertion of Nominalization	[*refusal*] [*by parents*] [*of x*]
5. Unspecified NP Deletion	[*refusal*] [*by parents*]
6. Subject Preposing	[*by parents*] [*refusal*]
7. Object Adjunction	—
8. Object Preposing	—
9. Subjective Genitive	[/GEN/ + *parents*] [*refusal*]
10. Subject Adjunction	[*parents refusal*]
11. Morph Adj	[*parental refusal*]

TABLE 5.2
Product Nominalization in an Objective NOM CN

Transformation	Output Structure
First cycle	/PRODUCT/ ## CRITICIZE *x* MUSIC ##
1. Lexical Insertion	/PRODUCT/ ## [*criticize*] [*x*] [*music*] ##
2. Case Prep Spelling	/PRODUCT/ ## [*criticize*] [*by x*] [*of music*] ##
Second cycle	
3. Product Nominalization	[/PRODUCT/ + *criticize*] [*by x*] [*of music*]
4. Lexical Insertion of Nominalization	[*critique*] [*by x*] [*of music*]
5. Unspecified NP Deletion	[*critique*] [*of music*]
6. Subject Preposing	—
7. Object Adjunction	[*critique* [*of music*]]
8. Object Preposing	[*music critique*]
9. Subjective Genitive	—
10. Subject Adjunction	—
11. Morph Adj	[*musical critique*]

provided in the sections to follow. The reader is again reminded that since the modifiers in these CNs are all derived from generic nouns, any specification of number in the forms shown in these tables is motivated by requirements of surface syntax (i.e., to make the forms easier to read) rather than by underlying semantics.

The absence of determiners in these derivations, on the other hand, is

TABLE 5.3
Act Nominalization in a Multi-Modifier NOM CN

Transformation	Output structure
First cycle	/ACT/ ## POLLUTE INDUSTRY WATER ##
1. Lexical Insertion	/ACT/ ## [*pollute*] [*industry*] [*water*] ##
2. Case Prep Spelling	/ACT/ ## [*pollute*] [*by industry*] [*of water*] ##
Second cycle	
3. Act Nominalization	[/ACT/ + *pollute*] [*by industry*] [*of water*]
4. Lexical Insertion of Nominalization	[*pollution*] [*by industry*] [*of water*]
5. Unspecified NP Deletion	—
6. Subject Preposing	[*by industry*] [*pollution*] [*of water*]
7. Object Adjunction	[*by industry*] [*pollution* [*of water*]]
8. Object Preposing	[*by industry*] [*water pollution*]
9. Subjective Genitive	[/GEN/ + *industry*] [*water pollution*]
10. Subject Adjunction	[*industry* [*water pollution*]]
11. Morph Adj	[*industrial* [*water pollution*]]

TABLE 5.4
Agent Nominalization in an Objective NOM CN

Transformation	Output structure
First cycle	*x* ## PLAN *x* CITIES ##
1. Lexical Insertion	*x* ## [*plan*] [*x*] [*cities*] ##
2. Case Prep Spelling	*x* ## [*plan*] [*by x*] [*of cities*] ##
Second cycle	
3. Agent Nominalization	[/AGENT/ + *plan*] [*of cities*]
4. Lexical Insertion of Nominalization	[*planner*] [*of cities*]
5. Unspecified NP Deletion	—
6. Subject Preposing	—
7. Object Adjunction	[*planner* [*of cities*]]
8. Object Preposing	[*cities planner*]
9. Subjective Genitive	—
10. Subject Adjunction	—
11. Morph Adj	[*urban planner*]

due simply to ignorance. There has been next to no research done in recent years on the sources of determiners in nominalizations; this neglect is the more surprising in view of the pivotal role (noted earlier) that nominalizations have played in the theoretical conflicts between the lexicalist and generative semantic schools of transformational grammar.[17]

[17] Coursaget-Colmerauer 1973 is the only article I know of which focuses on the deter-

Patient Nominalization in a Subjective NOM CN

Transformation	Output structure
First cycle	*y* ## APPOINT PRESIDENTS *y* ##
1. Lexical Insertion	*y* ## [*appoint*] [*presidents*] [*y*] ##
2. Case Prep Spelling	*y* ## [*appoint*] [*by presidents*] [*of y*] ##
Second cycle	
3. Patient Nominalization	[/**PATIENT**/ + *appoint*] [*by presidents*]
4. Lexical Insertion of Nominalization	[*appointee*] [*by presidents*]
5. Unspecified NP Deletion	—
6. Subject Preposing	[*by presidents*] [*appointee*]
7. Object Adjunction	—
8. Object Preposing	—
9. Subjective Genitive	[/**GEN**/ + *presidents*] [*appointee*]
10. Subject Adjunction	[*presidents appointee*]
11. Morph Adj	[*presidential appointee*]

Nonetheless, almost all discussion in the generative literature has centered solely on the nominalized head noun in such constructions (e.g., *proof, decision, solution, intelligence*), with the ubiquitous *John* occasionally but almost fortuitously moving up to fill a determiner slot (as in *John's proof, John's intelligence*). Prior to the publication of my own earlier studies (Levi 1973, 1974, 1975), no systematic derivations had been provided to account for even the smallest range of actually occurring nominalizations (such as those to be shown in [5.16]), nor had any of the major discussants (i.e., Chomsky, Fraser, Lakoff, McCawley, Newmeyer, Ross) published any studies that gave more than cursory attention to the variety of prenominal modifiers that can co-occur with these head nouns. It is as if the spotlight had been held on nominalizations just long enough to illuminate some features of the head noun (though by no means all), while the role of articles and other sister nodes of the head noun never emerged from the shadows.

One of the problems inherent in this subject stems from the fact that there may be as many as four principal NP nodes in the underlying structure (see [5.20] to follow), any one or all of which could have the semantic equivalent of a determiner associated with it. It is this multiplying factor of four that makes the analysis of such forms much more

---

miner constituent in NPs whose heads are nominalizations. Unfortunately, although the author raises a number of important questions concerning the surface distribution of determiners in the French nominalizations she is studying, she draws no trees and formulates no rules. Clearly much more work awaits the interested researcher in this area.

complex than that of determiners in single NPs (an area of the grammar that is far from fully understood in any case), since one must figure out just how the "determinant material" in each of the constituent NPs interacts with that of the others to produce a surface form. Variables such as number, referentiality, definiteness, and genericness will have to be taken into account, along with any other complications that may also be lurking about.

The absence of determiner nodes in the derivations to follow is thus unavoidable at this stage of our knowledge; as noted in the previous chapter, however, this lacuna implies that the overall adequacy of the theory cannot be fully judged or substantiated until the problems associated with determiners have been more satisfactorily investigated.

The transformations relevant to the derivations of NOM CNs are listed in (5.15), in their proposed order of application; orderings for which either intrinsic or extrinsic arguments may be advanced have been indicated in the standard manner.[18] Beginning with Section 5.3.2, each of these rules will be discussed in turn so that various aspects of importance to the overall analysis (such as arguments for ordering, problems in formulation, or possible alternative analyses) may be given more detailed attention.

(5.15)   **First cycle**
    1. Lexical Insertion (of V and arguments)   Obligatory
    2. Case Preposition Spelling   Obligatory

    Second cycle
    3. Nominalization   Optional
    4. Lexical Insertion (of head N)   Obligatory
    5. Unspecified NP Deletion   ??
    6. Subject Preposing   Optional
    7. Object Adjunction   Obl/Opt??
    8. Object Preposing   Optional
    9. Subjective Genitive Formation   Obligatory
    10. Subject Adjunction   Optional
    11. Morphological Adjectivalization   Optional
    12. *by* → *of* Adjustment   Optional

The perhaps surprising number of optional transformations included in these derivations is a reflection of the fact that NPs derived by nominali-

[18] Despite the fact that Transformations 11–12 are listed here under "Second cycle," the possibility remains open that this sort of spelling rule actually applies postcyclically. However, I have not yet found any evidence that might help decide the question. Note also that the *by* → *of* Adjustment Rule shown in (5.15) is not included in Tables 5.1–5.5 since it in fact only applies in cases where the underlying arguments are not preposed, and hence do not form CNs; see discussion of this rule in Section 5.3.12.

zation (whether CNs or not) exhibit a wide variety of structural forms. To illustrate this, (5.16) includes a sampling of the grammatical variations that can be produced from the same underlying structures as those of the CNs derived in Tables 5.1–5.5. The numbers in parentheses in (5.16) refer to those transformations listed in (5.15) which apply nonvacuously in the generation of each expression.

(5.16)   a. *refusal by parents* (1, 2, 3, 4, 5)
          b. *refusal of [< by] parents* (1, 2, 3, 4, 5, 12)
          c. *parents' refusal* (1, 2, 3, 4, 5, 6, 9)
          d. *parental refusal* (1, 2, 3, 4, 5, 6, 9, 10, 11)
          e. *critique of music* (1, 2, 3, 4, 5, 7)
          f. *music critique* (1, 2, 3, 4, 5, 7, 8)
          g. *musical critique* (1, 2, 3, 4, 5, 7, 8, 11)
          h. *pollution by industry of water* (1, 2, 3, 4)
          i. *pollution of water by industry* (1, 2, 3, 4, 7)
          j. *industry's pollution of water* (1, 2, 3, 4, 6, 7, 9)
          k. *industrial pollution of water* (1, 2, 3, 4, 6, 7, 9, 10, 11)
          l. *water pollution by industry* (1, 2, 3, 4, 7, 8)
          m. *industrial water pollution* (1, 2, 3, 4, 6, 7, 8, 9, 10, 11)
          n. *city planner* (1, 2, 3, 4, 7, 8)
          o. *urban planner* (1, 2, 3, 4, 7, 8, 11)
          p. *presidents' appointees* (1, 2, 3, 4, 6, 9)
          q. *appointees of [< by] presidents* (1, 2, 3, 4, 12)

Let us now examine the details of each of the transformations listed in (5.15).

5.3.2 Lexical Insertion

On the first cycle, the verb and NP arguments of the embedded sentence are lexicalized by lexical insertion rules (whose detailed formulation need not concern us here). It is important to note that Lexical Insertion is intrinsically ordered prior to Case Preposition Spelling since the case prepositions provided by the latter rule are governed by the specific lexical verb that has already been inserted. (Additional reasons for ordering Lexical Insertion early in the cycle were given in Section 4.2.1; see especially page 130.)

5.3.3 Case Preposition Spelling

Although "Preposition Spelling" is the term most commonly seen in the literature for the transformation that inserts those prepositions that regularly show up in nominalizations, this transformation is more accurately characterized as a "Case Preposition" spelling rule. This rule was correctly identified as a case marking rule by Lakoff and Peters (1969):

> At some point in their derivations, all object noun phrases take prepositions as a kind of case marking. When substantivization takes place, that is, when a verb is transformed into a noun, the preposition remains:

(125)	*the killing of the men*
(126)	**the killing the men*
(127)	*the killer of the men*
(128)	**the killer the men*

> However, if the verb is not transformed into a noun, the preposition is deleted by a postcyclic rule [p. 132].

Ross (1972b), on the other hand, in discussing "the rule that deletes prepositions after verbs," refers only to prepositions and not to "case markers," as may be seen here:

> It is unlikely that these prepositions can be argued to have appeared in remote structure. Rather, it is probable that underlying NPs become PPs during the course of a derivation, by the transformational insertion of a preposition [p. 123, footnote 39].

Ross is correct, however, in speaking of the assignment of these (case) prepositions to "underlying noun phrases" in general, rather than just to the "object noun phrases" that Lakoff and Peters mention, since we must account not only for the *of* preposition that appears before most (though not all) direct object NPs in nominalizations, but also for the *by* that appears before the subject NPs, as in (5.16a, h, i, l). A single transformation that inserts these case prepositions according to underlying order of the arguments could account for both the usual *by* and the usual *of* to mark subject and direct object NPs, respectively.

As for the claim that this rule must be viewed as a "case preposition" spelling rule, there is considerable evidence to indicate that later rules (including some in the derivations in this chapter) must be sensitive to whether the "preposition" is a case marker, inserted transformationally without independent lexical content of its own, or whether it is a descendant of an independent semantic predicate that happens to emerge as a preposition in English. For example, *investigation by senators* may be transformed into *senators'* (or *senatorial*) *investigation*, but *impeachment by spring* may not be similarly transformed into *spring impeachment*; the "case preposition" *by* which marks subjects may be deleted, but the lexical preposition *by* that expresses 'not later than' may not be deleted. Differences like this one in syntactic behavior between case-marking tokens of *by* (or various object markers) and homophonous prepositions having

  5 CN DERIVATIONS BY PREDICATE NOMINALIZATION

their own lexical entries demonstrate that "case prepositions" and "true" prepositions must be systematically distinguished at various points in the derivation **after** each has been inserted. In the present framework, this distinction is made whenever necessary by a global rule which in effect remembers the semantic or grammatical content of the individual lexical item.

A formulation of the Case Preposition Spelling Rule appears in (5.17). It is assumed here that the choice of case prepositions inserted by this rule is lexically governed by the previously inserted verb. (Cf. *the demand for contraceptives* versus *the supplying of contraceptives; the attack on the city* versus *the invasion of the city*.) Nonetheless, to simplify the presentation and to avoid extraneous issues, the trees in this chapter show only the **unmarked** choices of case prepositions for transitive English sentences, namely, *by* for subject NPs and *of* for direct object NPs. In addition, although this rule is assumed to apply regardless of the number of arguments in each S (cf. *growth of cities, refusal by parents of advice, transmission by officers to recruits of messages, migration by artisans from villages to cities*), the trees in this chapter are restricted to showing at most two arguments (subject and direct object) since these are the only two arguments which play a significant role in the derivation of CN forms in particular. Finally, the reader is asked to note that in the series of trees to follow, the optionality of an element is indicated by parentheses around that constituent in the **input** tree alone; the output trees therefore show only trees in which all optional elements are present.

(5.17)   Case Preposition Spelling

Input tree: 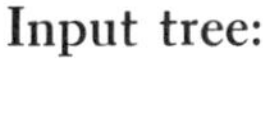 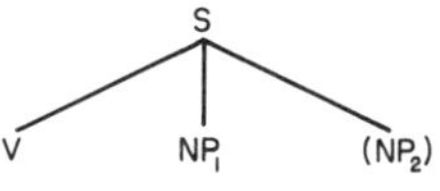

Output tree: 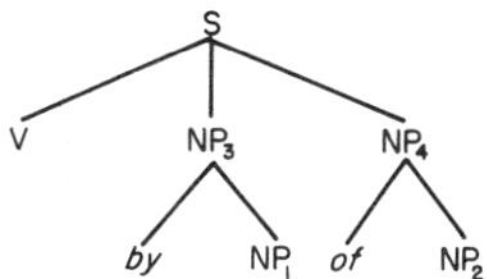

Comments
1. This rule applies obligatorily.
2. See text for comments on lexical government of this rule, as well as for restrictions imposed for pragmatic reasons on the preceding diagrams.

   This rule is intrinsically ordered prior to the Nominalization rules (see the following section), since the latter would remove the environment necessary for the application of the former; inasmuch as case prepositions can only be specified for the arguments of a **verb,** this specification must occur before the verb is incorporated into the category of Noun.

   The Case Preposition Spelling Rule has been written here so as to insert the case preposition *of* to mark an object NP. Although the choice of *of* as direct object marker is certainly the unmarked choice in English, there are in fact many verbs which require that their objects be marked with a case preposition distinct from *of.* A few examples of such verbs, together with their object markers, are given in (5.18)–(5.19).

(5.18)

Nonnominalized form	Nominalized form
He appeals **for** *mercy.*	*his appeal* **for** *mercy*
He depends **on** *his mama.*	*his dependence* **on** *his mama*
They report **on** *Cuba.*	*their report* **on** *Cuba*

(5.19)

Nonnominalized form	Nominalized form
She trusts her instinct.	*her trust* **in** *her instinct*
He opposed the increase.	*his opposition* **to** *the increase*
They attacked the city.	*their attack* **on** *the city*

In (5.18), verbs are shown whose objects are always case-marked; in (5.19), we see verbs whose objects are case-marked only within nominalizations. In both cases, however, the choice of the object marker is lexically governed by the specific verb. (I shall ignore here the possibility that many such choices are in fact governed by semantic classes of verbs, rather than by individual lexical verbs per se.)

   It is my claim that the object markers shown in (5.18)–(5.19) are inserted by precisely the same Case Preposition Spelling Rule as that shown in (5.17). That is, direct object case markers introduced by this rule are spelled *of* for verbs not otherwise marked, but are spelled *for, on, to, against,* and so forth for verbs that are correspondingly marked.[19] Wherever necessary in a derivation, global rules will "keep track of"

---

[19] See Levi 1975, Section 5.3 for a lengthier discussion of verbs whose direct objects are marked with case prepositions other than *of,* and for syntactic evidence supporting treatment of the prepositions in (5.18)–(5.19) as direct object markers.

whether the particular preposition is a grammatical case marker, or a lexicalization of a genuine predicate.

In this connection we may also note that, in general, *by* is selected as subject case marker only for subjects of transitive verbs, whereas subjects of intransitive verbs are normally marked with *of*; thus, we have *growth of cities* rather than **growth by cities*, and *death of seagulls* rather than **death by seagulls*. (Oddly enough, this pattern of case marking coincides precisely with that pattern which defines "ergativity" in other languages; see Sadock and Levi 1977 for further discussion of this point.) Here, too, global rules are needed to correctly interpret these tokens of *of* as subject markers, rather than as object markers, possessive genitive markers, or any of the many other functions that *of* may serve in English.

5.3.4 The Nominalization Rules

It is at this point in the derivation of NOM CNs that the appropriate rule of Nominalization, chosen from the four shown earlier in (5.11)–(5.14), may apply. These four rules are the crux of the process by which NPs with nominalization heads are derived in a generative semantic analysis like the present one.[20] Because of their theoretical significance, they have been accorded a section of their own (Section 5.2 just preceding) so that their analysis and justification could be presented in as much detail as our present knowledge permits.

We have seen, however, that many questions remain to be resolved in this area of the grammar. These include questions about the nature and number of the fundamental nominalization types, the choice of appropriate underlying structures for each, and the source of determiners in CNs with nominalizations as heads. In addition, the issue of the ordering of the Nominalization rules with respect to other rules such as Predicate Raising, and the question of their cyclicity or noncyclicity, remain matters of theoretical dispute. (Newmeyer [1976] has argued that both Predicate Raising and Nominalization must be precyclic rules; I will show in Section 5.4, however, that at least Nominalization must be regarded as requiring a cyclic application.) Nonetheless, we must begin our analysis somewhere, and it seems to me that the four rules of Nominalization proposed in Section 5.2 provide the most coherent and comprehensive foundation available to date within a generative framework.

Although not all the details of the rules of (5.11)–(5.14) need be reproduced here, we might profitably take a second look at those features

---

[20] Of course, in a lexicalist analysis like that of Chomsky 1970, nominalizations are not derived at all.

of the output structure that all of these rules have in common; these are illustrated by the diagram in (5.20). (The variable X is used here to represent any additional NP constituents derived from arguments of the underlying verb that would at this stage appear as sisters to the newly formed nominalization. The variable Y stands for whatever constituents, such as a determiner, might appear as left sisters to the nominalized verb at this point in the derivation.)

(5.20)

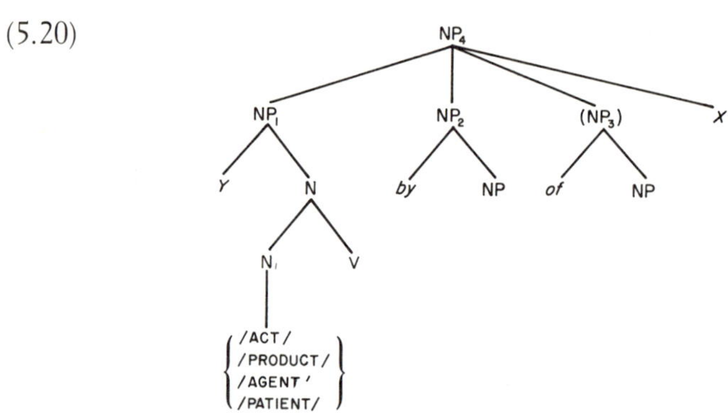

One of the major advantages of an analysis such as this one, whose Nominalization rules produce outputs like (5.20), is that it captures both the important similarities and the crucial differences among the four basic types of nominalizations. The similarities are reflected in the fact that all four rules produce an output structure whose leftmost NP constituent (*a*) dominates a noun that has incorporated the verb of the underlying S, and (*b*) stands in the same sisterhood relationship with the NP arguments from the underlying S as does the verb in the tree that is input to each rule. On the other hand, the crucial differences among the four basic types are reflected in the fact that the head noun combines the underlying verb with any one of four possible morphemes, each of which corresponds to a fundamentally different set of semantic and syntactic characteristics.

It is this latter aspect of the present analysis that can be used to account for at least some of the alleged idiosyncrasies of nominalizations that Chomsky (1970, 1972b) and others have used to criticize the descriptive adequacy and predictive power of **any** transformational analysis of these forms. Chomsky's rejection of a syntactic derivation of nominalizations (which he, oddly enough, refers to as "derived nominals") in favor of direct generation of these forms in the base (as NPs **without** component Ss) is motivated in large measure by the degree of semantic variation, allegedly to the point of idiosyncrasy, that he claims exists between a nominalization and its morphologically related verb (if indeed any such verb exists). Chomsky is certainly correct in drawing our attention to the fact that no analysis of nominalizations can ignore or deny the existence of many semantic and morphological irregularities in the data, but he is not justified in using such irregularities to reject out of hand all transformational derivations of these forms. As McCawley (1975) has pointed out in his review of Chomsky's major essay on nominalizations:

> The one thing about nominalizations which virtually all generative grammarians are agreed on is that the morphemic makeup and semantic content of nominalizations must be listed in the lexicon of a grammar. That conclusion, however, does not imply that in the relationship between nominalizations and their meanings all hell breaks loose [p. 218].

There is a double moral to the story being told here. First, a realistic perspective on the data demands that we neither deny the very real irregularities in the data nor exaggerate their intractability by any wholesale characterization of nominalizations as fundamentally rather than partially idiosyncratic. Second, we must recognize that whatever idiosyncrasies do exist in this part of the grammar pose just as serious a challenge to a lexicalist analysis as to a transformational one, and that neither lexicalists nor transformationalists can use these idiosyncrasies to reject the others' theory unless they can at the same time offer a competing analysis in which the allegedly idiosyncratic data emerge as consequences of generalizations captured (or capturable) within their analysis alone.[21] It is my claim here, however, that the formulation of the Nominalization rules shown in (5.11)–(5.14) permits us to achieve just this kind of theoretical progress.

To illustrate, let us focus on the fact that the Nominalization rules proposed here produce an output structure in which most features are shared but in which a four-way contrast is possible with respect to the morpheme that combines with the verb to determine the choice of the head noun appearing in surface structure; this four-way contrast among /ACT/, /PRODUCT/, /AGENT/, and /PATIENT/ permits us to make a wide variety of predictions about the morphological and semantic features of many nominalizations that might otherwise have to be characterized as unpredictable and idiosyncratic facts.

For example, we can account for both the morphological and seman-

---

[21] Since Chomsky's proposals on nominalizations (Chomsky 1970) do not meet this criterion, I find myself in agreement with such reviewers as McCawley (1975), Postal (1972), and Ross (1974a) who conclude that Chomsky has failed to present a cogent case for choosing a lexicalist treatment of these data over a transformationalist one.

tic differences between the nouns *amending* and *amendment* in (5.21) and (5.22) by tracing these differences to a distinction in the intermediate structures from which each noun is derived, with *amending* representing the appropriate lexicalization of a noun composed of /ACT/ + *amend*, and *amendment* corresponding instead to the noun composed of /PRODUCT/ + *amend*.

(5.21)
a. NOM: *The amending of the constitution*
b. CN: *The constitution(al) amending*
c. Non-NOM: *(The act of) amending the constitution*
d. Wrong NOM: **The amendment to the constitution*
e. Wrong non-NOM: **The product of (the act of) amending the constitution*

took seven years.

(5.22)
a. NOM: *The amendment to the constitution*
b. CN: *The constitution(al) amendment*
c. Non-NOM: *The product of (the act of) amending the constitution*
d. Wrong NOM: **The amending of the constitution*
e. Wrong non-NOM: **The act of amending the constitution*

was a piece of law that satisfied no one.

In an analysis such as the present one, we can predict the synonymity of the (a), (b), and (c) expressions in each of these two sets of data from the source structures that have been posited for Act and Product Nominalizations, respectively; in addition, the specification of /ACT/ and /PRODUCT/ as distinct head nouns in these structures permits us to predict the noninterchangeability of the two types of nominalizations in **any** of their surface manifestations, as demonstrated by the starred expressions in lines (d) and (e).

5.3.5 Lexical Insertion

Following the formation of a new head noun by one or another of the Nominalization rules, conditions for lexical insertion are met again; at this point, the bimorphemic head which is the typical output of all four Nominalization rules may be replaced by the specific noun that is listed in the lexicon as the appropriate lexicalization for that particular verb and nominalization type. For example, head nouns that contain /AGENT/ as

their first morpheme will be replaced in English by lexical items in which
that morpheme emerges as a suffix attached to the verbal stem (as in
planner, *supervisor*, *discussant*, and *analyst*) or by lexical items that are
superficially monomorphemic (as in *thief, cook,* and *pilot,* or *gauge, map,*
and *graph*); on the other hand, nominalizations containing /PRODUCT/ as
their first morpheme have no truly distinctive morphological expression
(i.e., an affix that marks Product Nominalizations alone) and will in most
cases emerge in forms that are homophonous with Act Nominalizations
(as in *error, comment, command, design,* and *analysis*).

As noted in the previous section, a complete characterization of the
grammar of nominalizations must recognize not only the syntactic and
semantic regularities which are, it is hoped, reflected in the Nominalization rules proposed here, but also the tremendous morphological variety
and semantic complexities that these forms display in English (and undoubtedly in many other languages as well). It is for this reason that
lexical insertion rules constitute an essential (if little understood) part of
the derivation of NOM CNs, since only rules of this sort provide access to
the idiosyncratic information stored in the lexicon and thus to the only
component in the grammar that is suited to catalog the many quirks and
kinks that appear in this domain. It is, then, the lexical insertion rules that
determine the choice of *decision* and not *decidal*, *refusal* and not *refu-
sion*, *stealer* in the context of baseball but *thief* elsewhere, and *analyst* for
humans but *analyzer* for machines.

Of course we could, if we wished, amuse ourselves at great length by
wandering among the many wild and seemingly uncivilized nominalization forms that have sprung up over the centuries within the more
disciplined ranks of the regularly patterning types, citing now this one and
now that one to show what a jungle it has all become. This course has
obviously been eschewed in the present work, since the indisputable
existence of **some** hopeless idiosyncrasies in no way diminishes the importance of the fact that nominalization processes in English **do** include
highly regular, productive patterns which native speakers systematically
exploit in both conventional and creative expression. It is patterns such as
these that the present derivations are intended to reflect, with the rules of
lexical insertion in question here serving as at least a partial bridge
between the syntactic and semantic regularities and the morphological
and semantic idiosyncrasies of these forms.

Unfortunately, precise details concerning the operation of this kind of
lexical insertion rule are not yet known since the process of lexical
insertion of complex nouns has been studied even less than that of
complex verbs. Despite the insightful attention paid to this question in
such works as Gruber 1967, Bach 1968, McCawley 1970, and Ross 1974a,

little progress has been made in following up the basically programmatic discussions provided by these authors. A more substantive account of the workings of such lexical insertion rules must therefore await subsequent research.

5.3.6 Subject Preposing

This optional transformation moves the subject NP from a position just after the verb, to a position in front of the verb, **without** changing its relations of sisterhood to the nominalization NP and the object NP. (This is in contrast to the Object Adjunction Transformation, which follows.) The full description of this rule appears in (5.23), where the nominalized verb serving as the head of NP_4 has been marked as N_v to assist the reader in following structural changes; this notation, like other subscripts introduced earlier for mnemonic purposes, plays no role in the operation of this or any other rule. The reader will also note that the output of this rule is a structure that may not surface without further change; in this case, the potential ungrammaticality of the output is due to the presence of the case marker *by* as the first lexical element in the NP.

(5.23)   Subject Preposing
     Example: *evaluation by students (of courses)* $\Rightarrow$
                    **by students evaluation (of courses)*

Input tree:

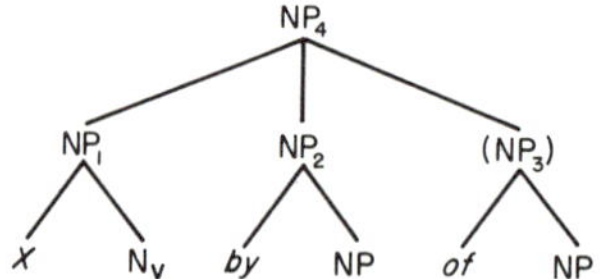

Output tree:

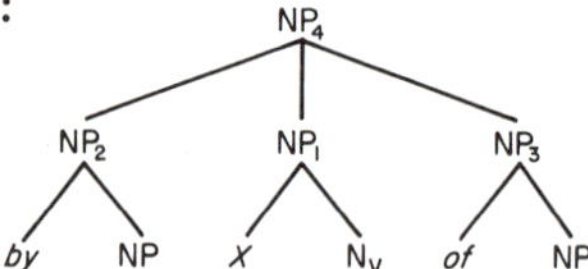

Condition
> *By* and *of* must be subject and direct object case markers, respectively.

Comments
1. This rule applies optionally.
2. Since the derivations of this chapter are designed to account for unmodified CNs only, the variable X shown in the preceding trees (which could theoretically include,

among other constituents, a determiner) will be assumed **for present purposes** to be empty. This pragmatically motivated assumption does not, however, correspond to any condition on the rule of Subject Preposing itself.

The adjunction of the subject NP to the higher NP (NP_4) rather than to the NP immediately dominating the nominalized verb (NP_1) is justified on the grounds that it permits the subsequent rule of Object Adjunction to produce a single NP constituent composed of only the nominalized verb and its object NP; this will be explained in more detail in the next section.

5.3.7 Object Adjunction

This transformation moves the direct object NP (shown here as NP_3) in a nominalization construction from its position as a sister of the NP dominating the nominalized verb (NP_1) to a position as its daughter, so that the underlying verb and its object become a single constituent, as shown in the tree diagrams of (5.24). The variables X and Y are included in these trees to show that this rule applies regardless of the presence or position of the subject NP; that is, Object Adjunction may take place whether the subject NP has been preposed by the prior but optional rule of Subject Preposing (in which case it would be represented by X in [5.24]), whether it is still in its original, postverb position (corresponding to Y below), or whether it has been deleted by Unspecified NP Deletion (in which case both X and Y could be empty, as in the derivation of a CN such as *water pollution* or *oil imports*)

(5.24)   Object Adjunction

    Example: *evaluation (by students) of courses* $\Rightarrow$
                 evaluation of courses (by students)

Input tree:

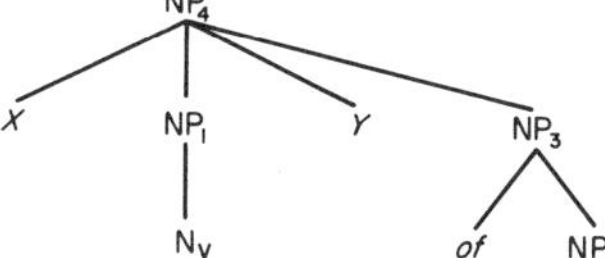

Output tree:

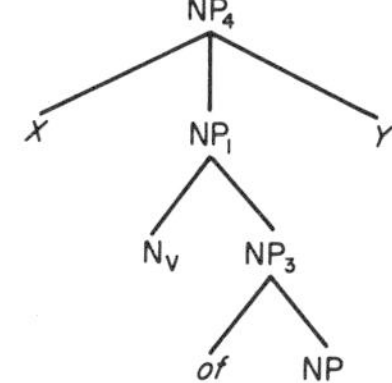

Condition

By and *of* must be subject and direct object case markers, respectively.

Comment

This rule appears to apply optionally when the input tree shows VSO order but obligatorily otherwise (i.e., when the subject NP$_2$ has been either preposed or deleted). See text for discussion.

The justification for this transformation is twofold. First, it is required to generate the set of nominalizations in which the underlying VSO order has been changed to VOS order, that is, in which the object NP precedes the subject NP without either having been preposed, as in *investigations of industry by senators* or *the destruction of the city by the enemy;* in a theory with underlying VSO order for English, this must be accomplished by a transformation. Second, it produces the correct constituent structures for such expressions, namely, with the nominalized verb and its object forming a single constituent.

The existence of this single constituent is argued for by two kinds of evidence. The first is based on pronominalization possibilities such as this one: *criticisms of Quine by Chomsky and those by Linsky* (where *those* = *criticisms of Quine*) as opposed to the unacceptable **criticisms by Chomsky of Quine and those of Hockett* (where *those* = *criticisms by Chomsky*). These examples (brought to my attention by James D. McCawley) clearly support the constituent structure indicated in the output tree of (5.24), and—not less significantly—the input structure of (5.23). The second source of support for the output tree of (5.24) comes from conjoined nominalizations whose common subject surfaces in prenominal genitival form, as in *the students' [criticisms of departmental programs] and [evaluation of individual teachers].* The correct analysis of shorter forms such as *students' evaluation of teachers* or *senators' investigations of industry* will therefore show a tree like that in (5.25a) rather than that in (5.25b), that is, a tree in which the subject NP is a sister of NP$_1$ rather than its daughter and in which the verb and its object form a single constituent.

(5.25)   a. **Right:**

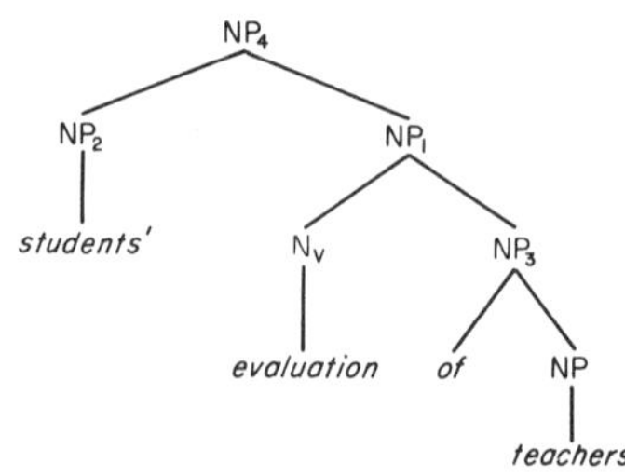

b. **Wrong:**

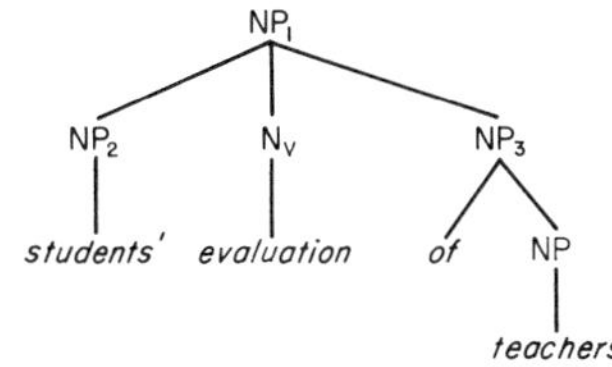

The data shown earlier in (5.16) indicate that this transformation (there numbered 7) applies to every nominalization with an object NP **except** for those where the underlying VSO order is preserved, as in (5.16h): *pollution by industry of water*. To generate all the data in (5.16), then, this transformation must be specified as applying obligatorily when the object NP is adjacent to the nominalized-verb NP (i.e., when the subject NP has been either preposed or deleted), but only optionally when the subject NP remains in its original, postverb position. I am uncertain as to just what theoretical consequences follow from imposing such a peculiar and unorthodox condition on this rule; nonetheless, it appears that only such a formulation will generate both the many possible word orders that are shown in (5.16) and their correct constituent structures.[22]

5.3.8 Object Preposing

In just those cases where both the nominalized verb and its underlying object are unmodified nouns (i.e., nouns without sister nodes), this transformation creates a new CN by Chomsky-adjoining the latter as a left sister of the former; structure-deleting conventions presumably then delete NP₃ and its two daughter nodes, including the case marker *of*. (We may note in passing that the preposing of the subject NP does not cause the deletion of the subject case marker *by*; since the node NP₂ is moved as a unit, no nodes remain behind for any structure-deleting conventions to remove. The subject case marker thus remains an integral part of NP₂, and is available to trigger later rules such as Subjective Genitive and *by* → *of* Adjustment. In contrast, the object case marker *of* is never moved into prenominal position; since the rule of Object Preposing does not prepose

---

[22] Another puzzle to be solved in this connection is why the VOS order produced by this rule seems to be stylistically favored over the original VSO order; for example, *recent investigations of industry stockpiling by concerned senators* seems to read more smoothly than the equivalent *recent investigations by concerned senators of industry stockpiling*, which has a certain choppiness to it; that this is not simply a function of length is shown by the fact that the shorter phrase *evaluation of teachers by students* (VOS) is also preferable to *evaluation by students of teachers* (VSO). The effect of this stylistic preference is to make this rule apply in almost but not quite all possible cases. I have no idea as to why this should be so.

the entire constituent NP_3—a constituent which includes the case marker *of*—but rather takes only the object noun to be a component of the new CN formed by Chomsky adjunction, the object case marker is not moved at all and so remains behind as one of the nodes that must be deleted subsequently.) The full description of this rule follows in (5.26); following the graphic convention introduced in Chapter 4, a box marks any N nodes that dominate a CN configuration.

(5.26) Object Preposing

Example: *evaluation of courses (by students)* $\Rightarrow$
course evaluation (by students)

Input tree:

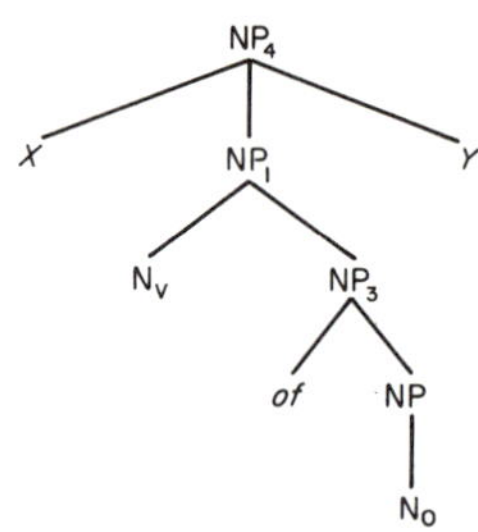

Output tree:

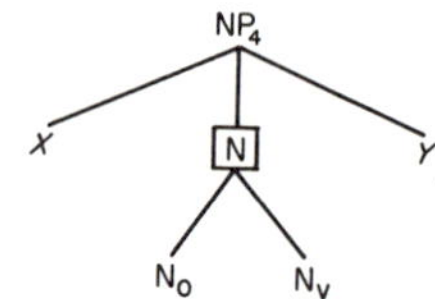

Conditions
1. *Of* must be an object case marker.
2. The node shown as N_o (for "object" noun) must have no sister nodes, and must be a proper or generic common noun.
3. NP_1 may dominate no other nodes but N_v and NP_3 (i.e., the nominalized verb may have no sister nodes besides the object NP_3).
4. X and Y are variables which may or may not include a case-marked subject NP (i.e., NP_2 in previous trees).

Comments
1. This rule applies optionally.
2. Structure-deleting conventions are assumed to be responsible for deleting NP_3 and its two daughter nodes after N_o has been Chomsky-adjoined to N_v.
3. This rule is one of two rules in the derivation of NOM

CNs (the other being Subject Adjunction) whose effect is to create a CN constituent. It is because of this fact that Conditions 2 and 3 must be imposed on the input tree.

Although the preceding formulation is written for the sake of simplicity as if it applied only when the object case marker is *of*, the rule is actually more general. Specifically, it can account not only for the transformation of *evaluation of teachers* and *investigations of industry* to *teacher evaluation* and *industry/industrial investigations*, respectively, but also for the generation of a wide variety of CNs whose underlying objects are marked with prepositions distinct from *of*, such as *monetary appeal* (< *appeal for money*), *age discrimination* (< *discrimination according to age*), *heart operation* (< *operation on the heart*), *alcoholic dependence* (< *dependence on alcohol*), and so forth. (Justification for analyzing these "prepositions" as object case markers may be found in Levi 1975, Section 5.3; for reasons of space, those arguments are not repeated here.)

Note that the constituent structure produced by this rule permits us to make two accurate predictions: first, that nouns or nominal adjectives derived from underlying object NPs will always appear immediately in front of the head noun, and second, that this descendant of the original object will be inseparable from the head noun once it has been moved to precede it. (The latter prediction is based on the assumption that lexical material dominated by a single node label of N—as are the two or more nouns that constitute a CN—form a unit whose components cannot undergo the rearrangements and distortions produced by the operation of syntactic rules on larger constituents.) The first prediction is borne out by forms like those in (5.27), and the second by the evidence in both (5.27) and (5.28), which together demonstrate that **no** prenominal modifiers (whether nouns, nominal adjectives derived from nonobject nouns, nonpredicating adjectives derived from underlying adverbs, or regular predicating adjectives) may be interposed between the object and head nouns of a well-formed CN.[23] (The previous sentence, in fact, contains an instance of the **only** way in which an object noun in a CN may be separated from its head, namely, when it appears in conjoined form with another object noun, as in *object and head nouns, air and water pollution*, or *stock and commodity purchases*.)

(5.27) a. $\begin{Bmatrix} \textit{industrial water} \\ \textit{*water industrial} \end{Bmatrix}$ *pollution*

---

[23] For a more detailed discussion of the ordering constraints in CNs with multiple nonpredicating modifiers, see Levi 1975, Chapter 8.

(5.27)

b. $\left\{\begin{array}{l}\textit{student monetary} \\ \textit{*monetary student}\end{array}\right\}$ *demands*

c. $\left\{\begin{array}{l}\textit{executive stock} \\ \textit{*stock executive}\end{array}\right\}$ *purchases*

d. $\left\{\begin{array}{l}\textit{American lunar} \\ \textit{*lunar American}\end{array}\right\}$ *probes*

e. $\left\{\begin{array}{l}\textit{monopoly price} \\ \textit{*price monopoly}\end{array}\right\}$ *fixing*

(5.28)

a. $\left\{\begin{array}{l}\textit{unavoidable water} \\ \textit{*water unavoidable}\end{array}\right\}$ *pollution*

b. $\left\{\begin{array}{l}\textit{exaggerated monetary} \\ \textit{*monetary exaggerated}\end{array}\right\}$ *demands*

c. $\left\{\begin{array}{l}\textit{alleged stock} \\ \textit{*stock alleged}\end{array}\right\}$ *purchases*

d. $\left\{\begin{array}{l}\textit{recent lunar} \\ \textit{*lunar recent}\end{array}\right\}$ *probes*

e. $\left\{\begin{array}{l}\textit{regular price} \\ \textit{*price regular}\end{array}\right\}$ *fixing*

5.3.9 Subjective Genitive Formation

This transformation is required in order to produce nominalizations in which the underlying subject NP appears in prenominal position with a genitive suffix attached to it; examples include *parents' refusals, senators' investigations, students' evaluations*, and *popes' appeals*. This rule, described in (5.29), has the effect of replacing an unambiguous subject marker (namely, the case marker *by*) with an ambiguous one (namely, the genitive morpheme shown here as /GEN/, which will surface in the form of the English "possessive /s/"). The rule therefore effects a change only in the preposed subject constituent but not in any other part of the nominalization construction.

(5.29)   Subjective Genitive Formation

Example: **by students evaluation of courses* $\Rightarrow$
students' evaluation of courses

Input tree:

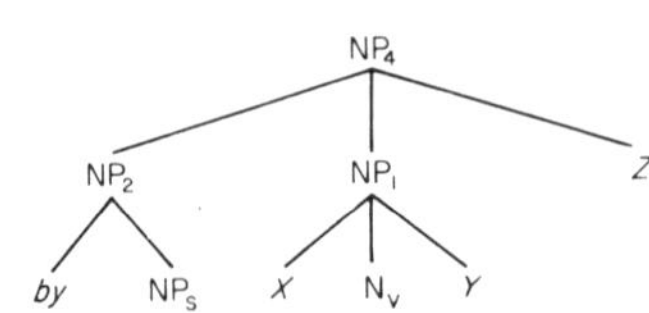

     5 CN DERIVATIONS BY PREDICATE NOMINALIZATION

Output tree:

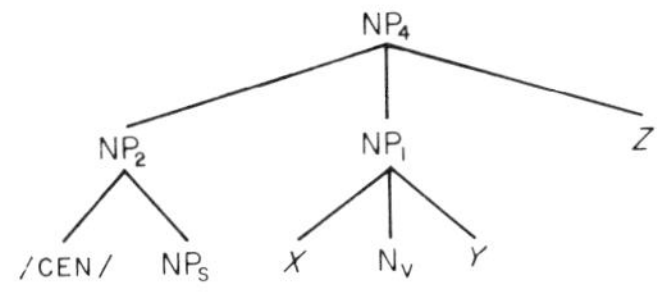

Condition

By must be a subject case marker.

Comments

1. This rule applies obligatorily.
2. The variables X and Y are included to show that the operation of this rule is not affected by the position (or prior deletion) of the underlying object NP.
3. /GEN/ represents the genitive morpheme lexicalized in English as the suffix *'s*.

The significance of this rule in regard to the grammar of NOM CNs is that it helps to create a derivational alternative to the formation of Subjective NOM CNs, since subject NPs which acquire genitival suffixes by this rule need not undergo the subsequent rule of Subject Adjunction, and it is only the latter rule that creates the Subjective NOM CN by adjoining an unmodified subject noun to the nominalized verb. Thus, underlying subjects of nominalized verbs which have been preposed (the first step necessary in the derivation of a Subjective NOM CN) may appear simply in the genitival form shown in the data of (5.30a), or they may undergo Subject Adjunction as well to emerge as prenominal modifiers within CNs, as shown in the data of (5.30b). It is, in fact, the existence of these parallel forms that demonstrates that Subject Adjunction must be an optional rule.

(5.30)   a. *parents' refusal*   b. *parental refusal*
            senators' investigations   *senatorial investigations*
            students' evaluations   *student evaluations*
            popes' appeals   *papal appeals*
            managers' decisions   *managerial decisions*

The fact that the forms in (5.30a), unlike those in (5.30b), are **not** CNs may be demonstrated by observing that the former permit the interposition of various modifiers between the subject and the head noun (as in *parents' adamant but ultimately counterproductive refusals* or *popes' urgent pre-Christmas appeals*), whereas the latter do not (cf. **parental adamant but ultimately counterproductive refusals* or **papal urgent pre-Christmas appeals*). There are therefore two reasons for considering the rule of Subjective Genitive Formation as a relevant but less than integral part of the CN formation process: First, its effect is to provide a non-CN alternative for the expression of subjects of nominalized verbs, and second, it applies to all NPs (as in *my lazy older sister's decision*) and not just to the unmodified nouns that we have seen are the required input for the other, central CN formation rules.

Before concluding this discussion of the Subjective Genitive Rule, we ought to inquire as to why this rule applies only to subject NPs, in view of the fact that genitival suffixes are also found on certain preposed object NPs, as indicated in (5.31).

(5.31)   a. *the city's destruction by the enemy*
            b. *the project's rejection by the committee*
            c. *the professor's repudiation by her protégé*
            d. *the child's examination by the nurse*

It is my contention that these preposed object NPs acquire their genitive suffix by a rule that is distinct from the Subjective Genitive Transformation described in the preceding paragraphs; this contention is supported by three pieces of evidence. First, the object NPs in (5.31) are systematically distinct from the object NPs in the nominalizations under study here in that the former are invariably definite and nongeneric, whereas the NPs in this study (exemplified by the sample in [5.16]) are invariably generic.[24] Second, the rules which form CNs are apparently sensitive to this distinction in definiteness and/or genericness since the only object nouns that may form CNs (in either an N–N or an Adj–N configuration) are those which are generic, like those in (5.32); in contrast, definite nongeneric objects, like those in (5.33), do not surface as part of Objective NOM CNs.

(5.32)   a. *destruction of cities*   = *city/urban destruction*
            b. *monitoring of hearts*   = *heart/cardiac monitoring*
            c. *control of vehicles*   = *vehicle/vehicular control*
            d. *spraying of trees*   = *tree/arboreal spraying*

(5.33)   a. *destruction of the city*   ≠ *the city/urban destruction*
            b. *repudiation of the professor* ≠ *the professor/professorial repudiation*

[24] As noted earlier, the only **common** nouns that may become part of a CN are generic nouns; it is true, however, that proper nouns may also form prenominal modifiers within NOM CNs, as in *Kennedy/Johnsonian appointments* or *Berlin/Parisian reconstruction*. Since, however, the contrast here between generic and nongeneric (and between definite and nondefinite) nouns manifests itself only among common nouns, my general practice of excluding proper nouns from this analysis will be continued here.

          5 CN DERIVATIONS BY PREDICATE NOMINALIZATION

c. *dismissal of the manager* ≠ *the manager/managerial dismissal*

d. *analysis of the play* ≠ *the play/dramatic analysis*

The reason that the NPs with postnominal **definite** objects shown on the left in (5.33) are not synonymous with the NPs on the right (containing CNs plus determiners) is that the latter have a clearly generic, nonspecific sense that the former necessarily lack; thus, *the urban destruction in Vietnam* can only be understood as 'the destruction of cities in Vietnam' (or, on another reading, 'the destruction in cities in Vietnam') but not as 'destruction of **the** city in Vietnam'. Finally, the difference between the two sets of object NPs is manifested in the very fact that the definite object NPs **can** appear with genitive suffixes, as illustrated in (5.31), whereas the generic ones may not, as shown in (5.34):

(5.34) a. { *city* / *urban* / **cities*' } *renewal*   b. { *vehicle* / *vehicular* / **vehicles*' } *control*

   c. { *heart* / *cardiac* / **hearts*' } *massage*   d. { *tree* / *arboreal* / **trees*' } *spraying*

I believe that these three facts offer sufficient syntactic evidence to support the claim that the Subjective Genitive Formation rule is indeed appropriately titled.

5.3.10 Subject Adjunction

By this point in the derivation, the direct object has formed a single constituent with the nominalized verb (though not necessarily in a CN configuration), but the subject has remained syntactically independent of the head noun. We will now see how the next rule in the derivation, that of Subject Adjunction, may alter this state of affairs in just those cases where the subject NP consists of an unmodified (i.e., sisterless) N that also meets the general requirement for CN formation that it be either a generic common or a proper noun.

As described in (5.35), this rule forms a new CN in the tree by Chomsky-adjoining the subject noun (shown here as N_s) as a left sister to the head noun of the larger construction, thereby forming a single constituent and conferring syntactic sisterhood on two formerly "unrelated" N nodes. As noted earlier in the discussion of (5.30), the existence of many nominal pairs such as *parents' refusal* and *parental refusal* demonstrates that this rule must be an optional one.

(5.35)   Subject Adjunction
          Examples: a. *students' evaluation (of courses)* $\Rightarrow$
                             student evaluation (of courses)
                          b. *students' course evaluations* $\Rightarrow$
                             student course evaluations

Input tree:

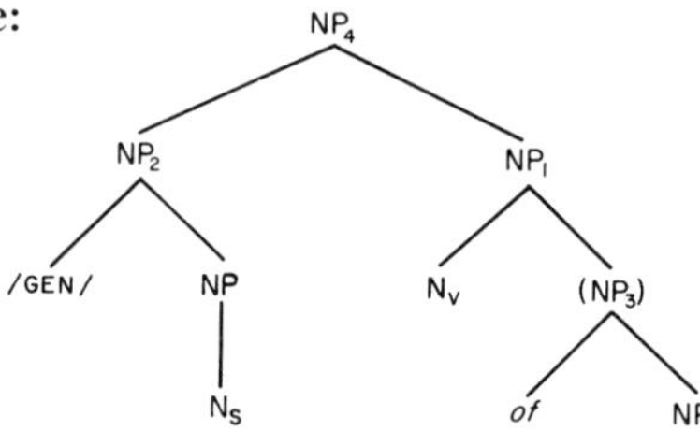

Output tree:

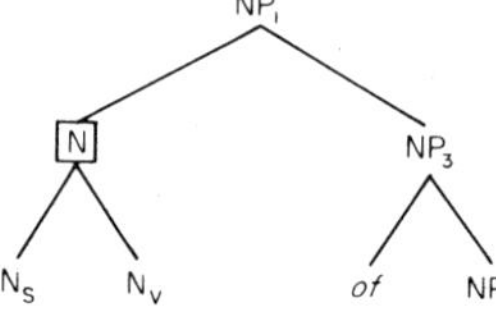

Conditions

1. *Of* represents the direct object case marker.
2. The node designated as N_s must have no sister nodes, and must be a proper or generic common noun.
3. The node designated as N_v must have no sister nodes besides (optionally) the case-marked object NP_3.

Comments

1. This rule applies optionally.
2. A tree-pruning convention removes NP_4, whereas other structure-deleting conventions are presumed to be responsible for deleting NP_2 and its two daughter nodes after N_s has been Chomsky-adjoined to N_v.
3. This rule is one of two rules in the derivation of NOM CNs (the other being Object Preposing) whose effect is to create a CN constituent (shown here as the boxed N). It is because of this that Conditions 2 and 3 are imposed.
4. In these trees, the node designated as N_v may dominate only the original nominalized verb (e.g., a noun such as *evaluation* or *control*) or it may dominate a CN formed from the nominalized verb and its object by the prior rule of Object Preposing (e.g., a CN such as *course evaluation*

or *pest control*). In the latter case, the input tree would not include NP_3.

The constituent shown as N_v in the preceding trees is so labeled for the convenience of the reader; the subscript, however, has no theoretical standing whatever. Thus, the structural description for this rule (represented in [5.35] by the input tree and the conditions) requires only a constituent labeled N to be present (with or without one NP sister) as a daughter of NP_1. If this N in fact dominates only the original nominalized verb, the subject N will be Chomsky-adjoined to it to create a CN like that shown as *student evaluation* in (5.36a); on the other hand, if this N happens to dominate a CN already formed by Object Preposing, the result will be another, more complex CN with a structure like that shown for *student course evaluation* in (5.36b). In either case, a new CN—and hence, a new constituent—will have been created, of which the subject noun will henceforth be an integral component.

(5.36)   a.

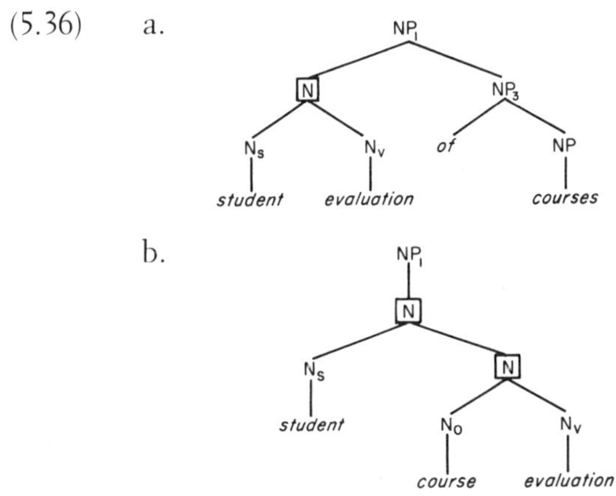

b.

In this context, readers are reminded that although the exposition here and throughout is primarily limited to the simplest cases of CN formation, the recursion which is a fundamental feature of the process means that **every** mention of an N node in these rules would be satisfied not only by the "simple" nouns used thus far, but also by any CN form **of whatever length and complexity** that might have been generated by the application of these rules on earlier cycles. In the case at hand, these rules could produce not only simple SOV forms like *student course evaluation* but also more complex ones such as *[college student] [night course] evaluation*, *[electrical engineering night school student] [computer science department syntactic analysis course] evaluation*, and of course many others whose invention is left to the reader's imagination.

Although it has not been judged necessary to emphasize this aspect of the rules in every section, it is nonetheless essential to keep in mind that it is the specification of these nodes as N nodes in each of these rules that is the true source of the immense productivity of the CN formation process.

Three additional remarks may be made about this rule of Subject Adjunction. We may note first of all that it is the ordering of this rule after Object Preposing, together with the specific structures that the two rules produce, which guarantees that nouns (or nominal adjectives) derived from underlying subjects will always precede those derived from underlying objects in CNs of the "multi-modifier" type shown earlier in (5.8). Second, it is interesting that this rule is by no means the only one to create the distinctive CN configuration by one or more binary adjunctions of individual nouns; in this chapter, we have seen that Object Preposing also has the effect of making new CNs, while the previous chapter demonstrated that each of the nine RDP Deletion rules is also a source of CNs. It is for this reason that no individual rule has been called "CN Formation" in this work.

The third remark concerns the fact that the application of Subject Adjunction, although characterized in (5.35) simply as "optional," is apparently conditioned by whether or not the subject NP has a human referent, since subject nouns referring to humans differ systematically from those referring to nonhumans in the forms they may assume when **preposed** within a nominalization. Thus, we observe in (5.37) that unmodified subject nouns referring to humans typically appear either with genitival suffixes or in adjectivalized form within a CN but for some reason are prohibited from expression within an N–N CN (i.e., in non-suffixed form). In contrast, the nonhuman subjects shown in (5.38) never appear with a Subjective Genitive suffix, but are always adjoined to the nominalized head noun as a prenominal modifier within a CN; there, however, they may appear in either nominal or adjectival form.

(5.37)   a. *Vatican Radio broadcasts all* $\left\{\begin{array}{l}popes'\\ {}^{*}pope\\ papal\end{array}\right\}$ *appeals.*

        b. $\left\{\begin{array}{l}Kings'\\ {}^{*}King\\ Royal\end{array}\right\}$ *orders are always obeyed instantly.*

        c. *Baker was curious about* $\left\{\begin{array}{l}presidents'\\ {}^{*}president\\ presidential\end{array}\right\}$ *inquiries into*

    the matter over the last fifteen years.

        d. *We were shocked by* $\left\{\begin{array}{l}people's\\ {}^{*}people\\ popular\end{array}\right\}$ *reactions to the news.*

(5.38) a. *The grant is for study of* $\begin{Bmatrix} {}^{*}\textit{birds'} \\ \textit{avian} \\ \textit{bird} \end{Bmatrix}$ *reproduction.*

b. *Parkway trees suffer greatly from* $\begin{Bmatrix} {}^{*}\textit{automobiles'} \\ \textit{automotive} \\ \textit{automobile} \end{Bmatrix}$ *emissions.*

c. *The rate of* $\begin{Bmatrix} {}^{*}\textit{cells'} \\ \textit{cellular} \\ \textit{cell} \end{Bmatrix}$ *decomposition varies widely.*

d. $\begin{Bmatrix} {}^{*}\textit{Hormones'} \\ \textit{Hormonal} \\ \textit{Hormone} \end{Bmatrix}$ *suppression of symptoms is a risky business.*

When these distributional facts are reinterpreted in terms of relevant rules, it appears that in order for human subject nouns to surface as prenominal modifiers within an NP whose head is a nominalized verb, those nouns must undergo **both** the rules of Subject Adjunction and Morphological Adjectivalization, or **neither** of these rules. In contrast, for nonhuman subject nouns to surface in the same role, they **must** undergo at least Subject Adjunction, and **may** undergo Morphological Adjectivalization as well.

The distributional patterns observable in these data are somewhat unexpected, and it is difficult to determine just which formulations and which ordering of the rules involved (namely, Subjective Genitive, Subject Adjunction, and Morph Adj) can most straightforwardly produce the desired results. For example, even though one might attempt to block the asterisked forms in (5.38) by placing a condition on the Subjective Genitive Formation rule which would exclude nonhuman subjects from undergoing the rule, there is no way in which the asterisks on the N–N forms in (5.37) could be so simply accounted for. In their case, either we must complicate the description of Morph Adj by saying that this normally optional rule applies obligatorily in just those cases where the first noun is derived from an underlying human subject, or we must impose some sort of output constraint to block the expression of these subjective nominalizations in N–N form.

Unfortunately, not only are all these alternatives inelegant and ad hoc, but none of them appears to work for 100% of the data. (Moreover, there appear to be dialectal or even idiolectal differences in judging acceptability of the forms in question.) The rule formulations and ordering shown in Sections 5.3.9, 5.3.10, and 5.3.11 therefore represent the most adequate analysis that I have been able to devise, but one that does not yet account for the full range of relevant data.[25]

---

[25] The possibility remains open, of course, that an approach focusing more on whatever

One final demonstration of the recalcitrant nature of the data is provided by the following examples, which represent exceptions to both of the general morphosyntactic patterns documented in (5.37) and (5.38). Thus, (5.39) shows human subjects that (atypically) escape the prohibition against appearing in an N–N CN, while (5.40) shows nonhuman subjects which (atypically) display the distribution that is normal for human subjects alone.

(5.39) a. $\left\{\begin{array}{l}\textit{Managers'}\\\textit{Manager}\\\textit{Managerial}\end{array}\right\}$ *attempts to increase productivity were duly rewarded.*

b. *The success of* $\left\{\begin{array}{l}\textit{parents'}\\\textit{parent}\\\textit{parental}\end{array}\right\}$ *intervention in crises depends crucially on the timing.*

(5.40) a. *In many cases,* $\left\{\begin{array}{l}\textit{nations'}\\\textit{*nation}\\\textit{national}\end{array}\right\}$ *exports are increased at the expense of domestic consumers.*

b. *Real estate agents show an irrational fear of the possibility of* $\left\{\begin{array}{l}\textit{cats'}\\\textit{*cat}\\\textit{feline}\end{array}\right\}$ *destruction of property.*

The facts we have been struggling with here can be summarized by the small chart in (5.41), showing the distribution that applies to the greatest number of cases. How the exceptions to this distribution should be handled, and why the English language should use just these morphosyntactic distinctions to distinguish among (most) human and nonhuman subjects, are questions for which answers are yet to be found.

(5.41)   Surface forms of preposed subject nouns

	N + GEN	N in CN	Adj in CN
+human	√	*	√
−human	*	√	√

---

functional or pragmatic distinctions can be associated with the forms in question might shed some useful light on the problem; however, no such study will be attempted here. (See Kay and Zimmer 1976 for a rather brief attempt at this sort of comparison.)

    5 CN DERIVATIONS BY PREDICATE NOMINALIZATION

5.3.11 Morphological Adjectivalization (Morph Adj)

This rule, discussed at length in Section 4.2.4 of the previous chapter, applies in precisely the same fashion to NOM CNs as to CNs derived by predicate deletion. Specifically, it applies (optionally) to left-branch nodes within CN forms, changing the noun to that nominal adjective which the lexicon specifies as its fully synonymous counterpart. The ordering of this optional rule must be either very late in the cycle or among the postcyclic rules; the latter is perhaps the more plausible, in view of the fact that I know of no rule (cyclic or not) that must be ordered **after** Morph Adj, for either NOM CNs or CNs derived by RDP Deletion. A few examples which illustrate the application of this rule to NOM CNs are shown in (5.42); others may be found in the data of (5.6)–(5.8) at the beginning of this chapter, and in Groups M–P in the Appendix.

(5.42)   *urban planning*   from   *city planning*
          cardiac massage   *heart massage*
          parental decision   *parent(s) decision*
          papal appeal   *pope appeal*
          dramatic criticism   *drama criticism*
          electrical generator   *electricity generator*
          acoustic amplifier   *sound amplifier*

This rule is quite properly considered an optional one, for the same reasons it was so designated in Chapter 4; those reasons will not be repeated here. Although one might claim that the asterisked forms shown earlier in (5.37) argue for a reformulation of Morph Adj as optional in some cases and obligatory in (these) others, we have seen that the nature of these data is so problematic that it is by no means clear that this is even a reasonable solution, much less an optimal one. Until the distribution of the forms in (5.37)–(5.40) is better understood, I will let stand the classification of Morph Adj among the many optional rules of these derivations.

5.3.12 by → of Adjustment

It is a curious fact of English that the subject NP of a nominalization which is normally marked with *by* (in postnominal position) may appear with *of* instead in just those cases where the direct object NP (itself marked by *of*) has been either preposed or deleted, as may be seen in (5.43).

(5.43)   a. *many refusals* $\begin{Bmatrix} by \\ of \end{Bmatrix}$ *the president*

b. *the shooting* $\left\{\begin{matrix}by\\of\end{matrix}\right\}$ *the hunters*

c. *investigations* $\left\{\begin{matrix}by\\of\end{matrix}\right\}$ *senators*

d. *investigations* $\left\{\begin{matrix}by\\*of\end{matrix}\right\}$ *senators of industry*

e. *investigations of industry* $\left\{\begin{matrix}by\\*of\end{matrix}\right\}$ *senators*

f. *industrial investigations* $\left\{\begin{matrix}by\\of\end{matrix}\right\}$ *senators*

The reader can easily see that the only grammatical expressions in which a subject NP originally marked with *by* may appear with *of* are those where the object has been deleted, as in (5.43a,b,c), or where it has been moved to precede the head noun, as in (5.43f). The constraint thus seems to be that only one *of* may appear in the environment following the nominalized verb.

That this is an accurate description of the facts is supported by data from two other kinds of nominalizations in which the same constraint appears to be operating. The first of these, nominalizations of intransitive verbs, is illustrated in (5.44).

(5.44)   a. *death* $\left\{\begin{matrix}*by\\of\end{matrix}\right\}$ *seagulls*

b. *falling* $\left\{\begin{matrix}*by\\of\end{matrix}\right\}$ *autumn leaves*

c. *arrival* $\left\{\begin{matrix}?by\\of\end{matrix}\right\}$ *dignitaries*

d. *growth* $\left\{\begin{matrix}?by\\of\end{matrix}\right\}$ *adolescents*

These data show that the subjects of nominalized intransitive verbs always **may** be marked with *of*, and in some cases **must** be so marked. (No determination of the precise conditions for marking these subjects with *of* will be made here, especially since even the limited data of [5.44] suggest that several factors may be involved.)

The second case, illustrated in (5.45), consists of verbs whose objects are invariably marked with a preposition distinct from *of*.

(5.45)   a. *an appeal* $\left\{\begin{matrix}by\\of\end{matrix}\right\}$ *the church for money*

b. *reports* $\left\{\begin{matrix}by\\of\end{matrix}\right\}$ *the leaders on pending negotiations*

c. *the preference* $\left\{\begin{array}{c}*by\\of\end{array}\right\}$ *parents for increased participation*

d. *the dependence* $\left\{\begin{array}{c}*by\\of\end{array}\right\}$ *students on last-minute cramming*

Here too we see that, in the absence of another *of* in the environment following the nominalized verb, the use of *of* to mark the subject NP is grammatical in all cases, and obligatory in some.

It thus appears that the curious fact cited at the beginning of this discussion is but a special case of a more general phenomenon (not the less curious for its increased generality) which restricts the case marker *of* to a single occurrence in the environment following a nominalized verb; moreover, the same constraint applies regardless of whether the *of* is an original case marker on a subject or object NP, or one derived by transformation from an agentive *by*.

We must therefore conclude that despite what Newmeyer (1970) has aptly called "the bizarreness of a *by* → *of* rule [p. 411]," such a rule must be included in the grammar of English in order to generate those cases in which both *by* and *of* are possible case markers for the same (subject) NP. (Cases which allow only *of*, such as [5.44a,b] and [5.45c,d], can be derived by marking the subjects with *of* to begin with.) One formulation of this optional rule appears here:

(5.46)   *by* → *of* Adjustment

Examples: a. *the shooting by the hunters* ⇒
             the shooting of the hunters

b. *reports by officials on pending legislation* ⇒
reports of officials on pending legislation

Input tree:

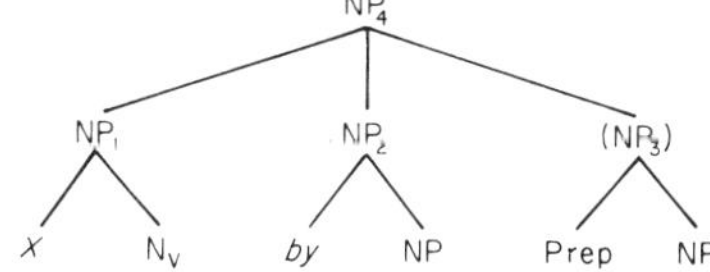

Output tree:

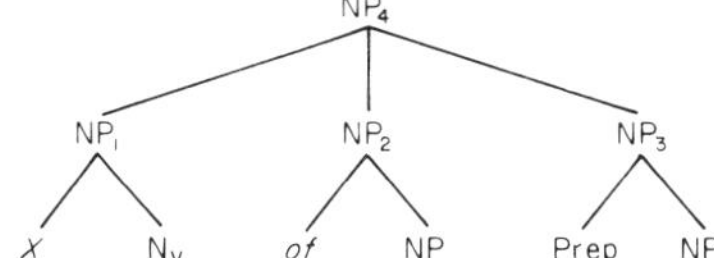

Condition

By must be a subject case marker.

Comment

This rule applies optionally, and possibly postcyclically.

Although a condition might have been imposed on this rule which would block its application in just those cases where the preposition following the subject NP would be *of*, I have assumed here instead that such negatively formulated conditions on transformations (i.e., conditions of the form "This rule applies except where . . ." or "Apply this rule unless . . .") are better stated as output constraints. In this kind of treatment, the rule of (5.46) applies optionally without conditions imposed on any case marker besides *by*; a surface structure (output) constraint is then used to rule out derivations in which the application of the rule happens to produce the unacceptable double-*of* sequence (i.e., *of* NP *of* NP). This approach provides a unified explanation (though not necessarily the only one possible) for the appearance of *of* in both the data of (5.43) where objects must be preposed or deleted, and the data of (5.45) where objects may remain unmoved. (One possible motivation for the existence in English of a mechanism to block the double-*of* sequence is the avoidance of case role ambiguity, since forms like *hiring of women of minorities* or *probes of the press of right-wing organizations* prevent us from ascertaining the original grammatical relations; on the other hand, the classic example of *the shooting of the hunters* reminds us that this condition may at best reduce case role ambiguities, but does not go so far as to eliminate them.)

Finally, the reader's attention is called to the fact that since this rule operates on NPs that have **not** been preposed within nominalizations, it is strictly speaking not a part of the CN formation process, since CNs are formed only from preposed arguments. This rule has been included nonetheless since it is relevant to the surface expression of a number of stylistic variants of the CN forms with which we are more centrally concerned.

5.4 Recursion and Cyclicity of NOM CN Derivations

In Chapter 4, I claimed that the manifest recursiveness of CNs formed by predicate deletion constitutes the strongest argument in favor of the cyclic application of the crucial rules enumerated there, especially the rules of Compound Adjective Formation and RDP Deletion. The set of CNs formed by predicate nominalization is no different in this respect, since the rules for forming NOM CNs can also operate recursively, producing CNs from CNs by repeated, cyclic applications. For example, the rules that combine to produce simple (i.e., one-cycle) Subjective NOM CNs like those in (5.6) may reapply to their own outputs to produce

     5 CN DERIVATIONS BY PREDICATE NOMINALIZATION

additional nominalizations, whose underlying subjects are themselves Subjective NOM CNs; examples of the outputs of such repeated applications are shown in (5.47), with brackets setting off the NOM CN which is both the output of the first cycle and the subject of the verb nominalized on the second cycle.

(5.47)   *[cellular decomposition] increase*
            [senatorial investigation] costs
            [spinal inflammation] spread
            [popular knowledge] explosion
            [automobile emission] poisoning
            [heartbeat] acceleration
            [cardiac arrest] recurrences
            [papal appeal] influence

Not surprisingly, the steps by which Objective NOM CNs are produced may also be repeated on a higher cycle to produce "double objective" nominalizations like the Product types shown in (5.48) and the Agent types shown in (5.49).

(5.48)   *[acoustic amplification] restrictions*
            [drama criticism] survey
            [fabric sale] advertisement
            [word formation] studies
            [water pollution] controls
            [space exploration] financing
            [predicate raising] justification

(5.49)   *[science teacher] evaluator*
            [air conditioner] repairer
            [urban planner] instructor
            [artistic manager] representative
            [life preserver] restorer
            [coal miner] supervisor
            [blood donor] recruiter
            [electrical generator] manufacturer

Additional support for the cyclicity of the derivations proposed in this chapter is derived from the fact that the rules outlined here interact with the demonstrably cyclic rules of Chapter 4 to produce CNs by derivations that involve both predicate deletion and predicate nominalization. Let us first consider cases in which the rules of Chapter 4 (to be called "the RDP rules") precede the rules of Chapter 5 (to be called "the NOM rules"), and then we will consider the reverse order of application.

In the former case, we would be interested in generating an expression like *suburban family recreational vehicle purchases* from the same NP-complement source as that which underlies this intermediate structure: *purchases by [families (who are) in the suburbs] of [vehicles (which are) for recreation]*; the essential feature here is that the subject and object NPs of the verb to be nominalized (here, *purchase*) must themselves be suitable sources for CNs of the predicate deletion type. The rules familiar from Chapter 4 would operate first, with IN Deletion producing the CN *suburban families* in subject position and FOR Deletion producing the CN *recreational vehicles* in object position. Then the rules of this chapter would operate in precisely the same manner as if the subject and object were not syntactically derived constituents, first nominalizing the verb and then preposing and adjoining each CN to the head noun so as to produce the surface form *[suburban family] [recreational vehicle] purchases* (exactly comparable to the simpler CN *family vehicle purchases*). Additional examples of nominalizations whose underlying arguments are themselves CNs of the predicate deletion type are provided in (5.50); in each case, the predicates deleted by CN formation from the subject and object NPs, respectively, are shown in parentheses on the right.

(5.50) *[minority student] [counseling service] demands* (BE, FOR)

     [gangster ring] [paper money] production (MAKE$_2$, BE)

     [funeral parlor] [floral wreath] purchases (FOR, MAKE$_2$)

     [state police] [tear gas] use (HAVE$_2$, CAUSE$_1$)

     [appliance store] [vacuum cleaner] repair (FOR, USE)

It is of course equally possible for a CN (or perhaps a single noun itself) first to be formed by the NOM rules of this chapter, and then, on a higher cycle, to serve as one of the nouns which form a CN derived by **subsequent** predicate deletion; here, the crucial rules of Chapter 5 would precede those of Chapter 4. The simplest cases would be those where just one of the two component nouns of a CN derived by predicate deletion was itself a nominalized verb (appearing, in these cases, without accompanying NP arguments); the NOM rules would produce the nominalization on an earlier cycle, after which the RDP rules would use it on a higher cycle as input to CN formation by predicate deletion. Examples of this sort, derived by deletion of various RDPs, are given in (5.51); the CNs in (a) all have nominalizations as head nouns, while the CNs in (b) show nominalizations serving as prenominal modifiers of (in most cases) nonnominalized head nouns. (Even in those cases where both nouns of the CN are nominalizations, as in *shock treatment* or *abortion vote*, the surface CN is still derived by predicate deletion rules operating **after** the formation of the two individually nominalized head nouns.)

(5.51) a. CAUSE₁: *mortal wound*
            HAVE₁: *pictorial report, musical production*
            USE: *hormonal treatment, vacuum abortion*
            BE: *parental visitors, avian marauders*
            IN: *elbow inflammation, urban riot*
            FOR: *sanitary engineering, equipment charges*

        b. USE: *shock treatment, talking cure*
            BE: *induction process, olfactory sense*
            IN: *translation problems, marketing controversies*
            FOR: *treatment room, communication system*
            ABOUT: *abortion vote, divorce law*

In more complicated cases, fully formed NOM CNs would constitute the input to the predicate-deletion derivation; such cases would produce complex CNs like those shown in (5.52) and (5.53). In the former, NOM CNs of various types serve as prenominal modifiers, whereas in the latter they are themselves the head nouns; the symbols in parentheses denote the kind of nominalization formed on the earlier cycle, followed by the specific RDP deleted on the later one.

(5.52) *[industrial growth] patterns*   (Subj-Act/Prod; IN)
        [parent participation] programs   (Subj-Act/Prod; HAVE₁)
        [national defense] budget   (Obj-Act; FOR)
        [bomb testing] fallout   (Obj-Act; CAUSE₂)
        [science teacher] institute   (Obj-Agent; FOR)
        [medical student] volunteers   (Obj-Agent; BE)

(5.53) *home [solar heating]*   (Subj-Act/Prod; IN/FOR)
        canine [heartbeat]   (Subj-Act/Prod; IN/HAVE₂)
        hormonal [symptom   (Obj-Act; USE)
           suppression]
        diabetes [blood test]   (Obj-Act/Prod; FOR)
        women [coal miners]   (Obj-Agent; BE)
        electric [can opener]   (Obj-Agent; USE)

Clearly there can be no doubt about the recursive nature of the rules that produce NOM CNs, in view of the complexity and variety of CN forms that appear as their output, either as the result of their application alone or as a result of their interaction with the RDP rules of Chapter 4.

Moreover, there simply is no way to account for this potentially unlimited recursion except by the cyclic application of the first 10 rules of the derivations shown earlier in this chapter.[26]

Not incidentally, the evidence presented here demonstrates that Newmeyer's arguments for the precyclicity of nominalization processes (Newmeyer 1976) are theoretically untenable. It is of some interest to note, however, that Newmeyer's proposals were apparently motivated in large measure by a desire to demonstrate that incorporation rules of the sort represented by nominalization, predicate raising, and "compound formation" (the latter explicitly analyzed in Newmeyer 1975) are of a fundamentally different nature than other syntactic rules (especially movement rules such as Passive, Subject Raising, and Dative Movement) which are demonstrably cyclic. Although it is now clear that this objective cannot be accomplished in terms of a precyclic analysis, we should not thereby be prevented from recognizing that the distinction Newmeyer sought to express is a genuine and an important one.

It is more likely, however, that the essential quality of this distinction is traceable not to a difference in derivational scheduling, but more directly to the fact that the incorporation rules in question are fundamentally **word formation** rules, whereas the other cyclic rules are fundamentally **sentence formation** rules. That this distinction is an essential one for the organization of a natural language is suggested by the fact that it is readily expressible in both syntactic and functional terms. Syntactically, we may distinguish between rules that apply within a constituent whose highest node is NP (as in the case of the RDP and Nominalization rules in this work) and those whose domain of application is a constituent whose highest node is S (as in the case of rules such as Passive and Dative Movement).[27] In functional terms, we might express the crucial distinction by contrasting those processes that create "reusable units" (of which the paradigmatic type is the lexical item, but which must also comprise nonlexicalized derived forms like novel CNs) and those which do not; alternatively, we might contrast processes that create names (a function

---

[26] As for the last two rules listed in these derivations, Morph Adj may yet be shown to apply postcyclically, while the *by* → *of* Adjustment rule does not, strictly speaking, even apply to CNs directly. Moreover, the changes that these rules effect make them seem more like superficial "spelling" rules than like the clearly syntactic rules that precede them, especially in view of the fact that no syntactic rule appears to be affected one way or another by their application or nonapplication. There therefore seems to be no compelling reason to argue for their inclusion within the cyclic set.

[27] Some refinement of this distinction is obviously necessary if Predicate Raising is to be linked with the first group rather than the second. Note also that the distinction sketched here is not necessarily to be equated with that between an NP cycle and an S cycle, although the two distinctions may well be related; see footnote 31 in Chapter 4 on this subject.

discussed in some detail in Section 3.3.2) and those that carry out more "sentential" functions. Note that the function of naming can be logically ascribed not only to those rules that create syntactic nouns (such as the RDP and Nominalization rules of this book) but to the verb-creating rule of Predicate Raising as well, inasmuch as the complex verbs which the latter produces may be appropriately viewed as names for events like changes in state (as in *redden* from BECOME + RED) and/or for actions (as in *deny* from SAY + NEG).

Although the preceding remarks are only speculative, the fact that the distinction in question permits us to form two apparently natural classes of rules within either a strictly syntactic or a functional analysis suggests that this is indeed one of the most fundamental distinctions that must be drawn—somehow—in any characterization of natural language.

6

IMPLICATIONS
OF THE THEORY

Now that the presentation of my theory is substantially complete, we may turn our attention to some of the related issues for which this study has relevance. In this chapter, a number of ramifications of the theory are presented, not so much to offer extensive elaborations of each topic, but rather to call the reader's attention to some of the broader implications of this work.

In the first section (6.1), I take up the influence of stylistic factors on the operation of the Morphological Adjectivalization rule proposed earlier. Implications of this study for theories of the lexicon are then sketched in Section 6.2, while the following section (6.3) deals with the ways in which syntactic, semantic, and pragmatic considerations interact in our production and interpretation of CN expressions. Finally, in Section 6.4, I return to the issues raised earlier concerning the universality of the RDP set and of CN formation in general, and attempt to separate the universal aspects of this process from the language-specific variations that have been observed.

6.1 A Note on Style

The appreciable number of optional rules that play a role in the derivation of CNs of both major types means that English speakers have numerous options in selecting the surface expression of the underlying forms we have considered here. For example, they may choose between nominalizations with postnominal and prenominal arguments, between the more compressed CN form and a fuller, more explicit periphrastic equivalent, and finally between a CN form whose prenominal modifier is a noun and one that contains a nominal adjective instead. It is this latter option that will receive explicit attention in this section.

The rule of Morphological Adjectivalization has been characterized throughout this work as an optional one. This characterization is strongly supported by the existence of CN pairs like those in (6.1), whose members differ solely in whether or not Morph Adj has applied to the prenominal element; as expected, the two forms of each pair are fully synonymous and, more interestingly, are apparently used in free variation.

(6.1) a. *industry unrest*   b. *industrial unrest*
     hand signal   *manual signal*
     parent figures   *parental figures*
     drama studies   *dramatic studies*
     language difficulties   *linguistic difficulties*
     sea turtles   *marine turtles*
     ocean winds   *oceanic winds*

On the other hand, there are also many cases of CNs where one of the two possible surface expressions is so highly favored in common speech that the alternative form is intuitively experienced as "jarring" at best, and "impossible" at worst; examples of this sort of unequal pairing are shown in (6.2) and (6.3), with the preferred variant appearing in the (a) column in each case.

(6.2) N–N preferred over Adj–N

    a. *picture book*   b. *pictorial book*
     eye infection   *ocular infection*
     father figure   *paternal figure*
     race riot   *racial riot*
     mind reader   *mental reader*
     construction materials *constructional materials*

(6.3) Adj–N preferred over N–N

    a. *logical fallacy*   b. *logic fallacy*
     digestive system   *digestion system*
     mental disturbance   *mind disturbance*
     racial disorder   *race disorder*
     instructional materials *instruction materials*
     maternal instinct   *mother instinct*

Despite the strong preferences that native speakers have for the forms in the (a) columns to those in the (b) columns, we must recognize that **all** of the forms in (6.2)–(6.3) are fully grammatical surface expressions which the rules described in the preceding chapters will generate in a perfectly straightforward fashion, and that the only difference of **grammatical** significance between the two sets of forms is that those in (6.2b) and (6.3a) have had the Morph Adj rule apply in the course of their derivation, whereas the others have not. It is for this reason that the two sets have been distinguished by labeling the (a) forms as "preferred" rather than by labeling the (b) forms as "ungrammatical."

Nonetheless, in view of the apparently unpredictable distributions of nominal adjectives such as those shown here, one may quite reasonably begin to wonder whether the Morph Adj rule is applied by native speakers in a totally capricious manner, or whether there are some discernible systematicities which govern, at least in part, the appearance of nominal adjectives within CNs. Although Bolinger (1967) seems to take the former position when he writes in regard to this issue that "word formation is a transformational wilderness [p. 31]," I believe that the evidence points more strongly toward the latter position. Thus, if this area of the grammar has any "wild" aspects to it, as Bolinger not unreasonably suggests, they derive not from the grammar itself but from the whims of humanity and the accidents of linguistic history; whatever "wilderness areas" exist are then not transformational in character but more of a morphological (and hence superficial) nature.

It is true that in many cases, nothing less arbitrary than historical accident appears to have influenced the choice between N–N and Adj–N versions of the same CN. For example, although *father figure* and *paternal figure* are both grammatical and synonymous, only the former has gained great currency in American speech; it does appear, however, that this choice must be attributed to historical and/or social reasons (if not pure chance) since there simply are no grammatical reasons to prefer either one over the other. Similarly, the fact that we ask our druggist for *nasal mist* (rather than *nose mist*) but for *nose drops* (rather than *nasal drops*) cannot be a function of the relative grammaticality of the contrasting CNs (since all four are equally and fully grammatical) but simply a function of the greater currency that one member of each pair has gained over the other (undoubtedly influenced in this case by repetition in advertising and marketing). It seems to me that the choices made within such pairs are as unpredictable—and as unamenable to linguistic analysis—as the length of skirts in next year's fashions or the next slang adjective to be adopted or coined by adolescents as a term of general approbation.

On the other hand, there is fairly consistent evidence in the data of somewhat more regular patterns appearing as the result of stylistic and/or sociolinguistic factors which influence the choice among various surface expressions, sometimes increasing and sometimes decreasing the likelihood that Morph Adj will be applied; although no absolute predictions can be made in terms of these factors, still the distribution of nominal adjectives in these cases is not wholly arbitrary.

It is an incontrovertible fact, easily observed and amply supported by linguistic analysis, that every native speaker has command of a number of stylistic levels of discourse, so that our choice of words will vary significantly depending on (among other things) the degree of formality inherent in a situation, the image we wish to project of ourselves within a given context, and judgments we have made of our addressee's background in terms of such variables as social class, education, ethnic identification, age, and occupation.[1] It is my claim that the frequency of application of the Morph Adj rule is one of the identifiable variables by which we can distinguish these stylistic levels, at least in those cases where the language provides a valid choice between a noun and an adjectival equivalent.

In English, it appears to be the case that the frequency with which nominal adjectives replace underlying nouns is positively correlated with the "level" of speech: The higher the stylistic or technical level of the discourse, and the higher the educational level of the speaker (and the addressee), the more likely that Morph Adj will have been applied to produce a nominal adjective. For example, medical personnel would be more likely than laymen to speak among themselves of *renal disease* rather than *kidney disease*, just as engineers or geologists would be more likely than nonspecialists to use *fluvial currents* in place of *river currents*, even though both members of each CN pair must be recognized as fully grammatical and semantically equivalent.

Of course, in many cases a nominal adjective that is used with relative frequency in the jargon of one particular profession may simply not be present in the lexicon of someone outside of that profession (although that person may belong to a different trade having its own esoteric jargon), regardless of the latter's level of education. This is certainly true of innumerable adjectives in any English dictionary, which even highly educated speakers may never have encountered. This simply means that the richness of one's vocabulary must be seen as a function, not only of one's "level" of education described along the customary vertical scale (as in the phrase *highly educated*), but also of one's "breadth" of education as measured along a horizontal scale.

[1] For more extensive discussion and documentation of this phenomenon from a variety of viewpoints, see Crystal and Davy 1969, Joos 1961, Labov 1972 (especially Chapter 3), or Turner 1973.

    6 IMPLICATIONS OF THE THEORY

On the other hand, there are certainly many nominal adjectives which **are** present in a lay person's vocabulary but which are unconsciously reserved for use in speaking at "higher" levels of discourse. For example, a police officer who might describe a chase scene to a friend in words like these, "We had to stop the chase because a car got in the way," might write for the official report, "Pursuit of the suspect was interrupted by a vehicular obstruction in the officers' path." Both *car* and *vehicular* would be in the officer's lexicon, but the latter would be used only for the more formal style that is presumably required in a written report to one's superiors.

It can also happen that the higher stylistic level conveyed by use of nominal adjectives will be exploited by a speaker even in informal situations, if that person judges it psychologically useful; this can be done for such purposes as showing off one's membership in some exclusive (jargon-defined) group, trying to demonstrate intellectual superiority over one's addressee, or (less obnoxiously) sharing the pleasures of linguistic creativity.[2] The stylistic flexibility that the Morph Adj rule sets up in English thus gives us a very useful tool for expressing in linguistic form a variety of social variables and psychological attitudes.

One of the reasons that this correlation exists in English between stylistic level and frequency of nominal adjectives is that the latter are almost always formed by the combination of a Latin or Greek nominal stem with a Latin or Greek adjectivalizing suffix, and such Classical words (whether nouns, verbs, or adjectives) have acquired a more prestigious or learned association than have synonymous words derived from non-Classical languages. Thus, *subterranean* somehow sounds more impressive than *underground*, and *aquatic insects* sounds more learned than *water bugs*. This distinction between "refined" and "ordinary" lexical equivalents is clearly part of the competence of native speakers of English, and represents one of the ways in which we can vary (consciously or otherwise) both linguistic and extralinguistic components of our message.

Since, however, this particular correlation between linguistic form and stylistic level is very much a function of the somewhat schizophrenic nature of English linguistic history (i.e., reflecting the large contributions made by both Germanic and Romance sources), we certainly cannot

---

[2] A case that may reflect all three purposes is that of the English professor in Tom Stoppard's play *Jumpers* who responds to his wife's screams of "Murder—Rape—Wolves!" from a neighboring bedroom with this shout: "Dorothy, I will not have my work interrupted by these gratuitous acts of lupine delinquency! [cited by Stanley Kauffmann, *New Republic*, 18 May 1974:18]" Stoppard has neatly captured here the sort of relentless exploitation of the Morph Adj rule that characterizes the speech of certain individuals within academic circles in particular.

expect that other languages, possessing idiosyncratic histories of their own, would necessarily manifest the same direct correlation between more formal speech and frequency of nominal adjectives. To begin with, not all languages have the same variety of surface syntactic options that English shows for expression of the same underlying form (cf. *decisions by parents, decisions of parents, parents' decisions, parent decisions, parental decisions*, for NOM CNs alone). Some languages simply do not permit a CN-like adjunction of two nouns without an additional marker (e.g., Japanese and, for the most part, French), others are relatively poor in adjectivalizing affixes (e.g., Akkadian and German, especially by comparison with Romance languages), whereas still others have morphosyntactic distinctions that are different from, though not necessarily more varied than, those available in English (e.g., Turkish, in which the functional and semantic equivalents of English CNs may appear with a "genitive" suffix added to each of two adjoined nouns, to the N–N collocation as a whole, or to neither of the two nouns).

In addition to crosslinguistic differences in the number and form of the surface variants available to express CN forms (or their equivalents), languages can also be expected to differ in the stylistic values attached to each of the options. For example, In Modern Hebrew N–N forms (known as *smixut* or "nouns in construct") are considered stylistically superior to other equivalent constructions (including, and perhaps especially, that of a noun modified by a nominal adjective); although this is essentially the reverse of the English pattern, the contributing factors are quite comparable, namely, the language family to which Hebrew belongs (in which N–N constructions happen to have an old and venerated place) and the highly idiosyncratic history of the modern-day language (involving, in the case of Hebrew, conscious and deliberate reconstruction from earlier historical stages of the language—a linguistic history quite unlike the unplanned and essentially unmanipulated confluence of Romance and Germanic components in the development of Modern English).[3]

It was noted earlier in Section 4.2.4 that while the **existence** of a given nominal adjective in the lexicon of English is closely tied to the etymology of the noun from which it would be derived, the **appearance** of such an adjective within a CN is dependent on numerous other factors. In this section, I have argued that these factors cannot be said to include syntactic or semantic considerations per se, but rather must be viewed as a combination of (*a*) historically determined but arbitrary distribution of forms (as suggested by the data in [6.2]–[6.3] in particular), (*b*) language-

---

[3] For further discussion of this difference between Modern Hebrew and English, see Levi 1976:27–28.

specific stylistic distinctions, assigned fairly systematically, which contrast the Adj–N and N–N variants of a particular CN, and (*c*) highly context-sensitive factors of a social and psychological nature. As a result of the last two factors in particular, the Morph Adj transformation may quite properly be viewed as a rule whose optionality provides the English language with a degree of stylistic flexibility which it might not otherwise enjoy.

6.2 The Content and Organization of the Lexicon

The research that has been reported here carries with it a number of implications concerning the content and organization of the lexicon; the two specific issues to be addressed in this section involve the listing of CNs as units in a lexicon or dictionary, and appropriate mechanisms for linking nominal adjectives with their underlying nouns. I will first review briefly some current practices in published monolingual dictionaries as an example of how **not** to approach these issues. The remainder of the discussion will then be framed in more theoretical terms, that is, in terms of structuring the lexicon as one component of a theoretical grammar; nonetheless, it should be fairly clear from the recommendations made in this theoretical context what the corresponding implications are for the construction of practical dictionaries.

An examination of some current dictionaries in the light of the preceding chapters reveals that certain lexicographic policies and practices have the unfortunate effect of providing information that is inconsistent, redundant, incomplete, and/or incorrect. Taking the *Oxford English Dictionary* (1971) as an example, I must note in all fairness that the difficulties involved in deciding which CN forms to list and which to omit are at least partially recognized in the section on "General Explanations" where the editors comment:

> There is thus considerable difficulty in determining to what extent combinations [= various compound forms and frozen expressions] are matters for the lexicographer, and to what extent they are merely grammatical. While no attempt is made fully to solve this difficulty, combinations formal and virtual are, for practical purposes, divided into three classes [p. xii].[4]

---

[4] The three classes in question are highly specialized CNs which are given full entries of their own, slightly specialized CNs for which only truncated definitions are provided, and CNs that are simply listed without comment within the entry for their N_1. See discussion to follow for examples.

In view of these frankly acknowledged difficulties, the present discussion may be taken in part as an attempt to assist practical lexicographers, as well as lexical theorists, to make the pertinent distinctions in a more coherent and consistent manner than has yet been achieved.[5]

One of the ways in which the *OED* implicitly acknowledges that some "combinations" are "merely grammatical" is by listing CN forms in an entry without specific definitions, usually under a separate heading of *attrib. and Comb.* (= attributive [use] and Combinations). For example, CN forms such as *horse-mane, horse-cart, horse-breeder, horse-face,* and over a hundred more are found within the entry for *horse;* they appear there without definitions but grouped in a fairly informative if somewhat inconsistent manner under such headings as "appositive" (*horse-beast*), "objective and objective genitive" (*horse-breeder, horse-breeding*), "for a horse" (*horse-feed, horse-ferry*), and "carried, drawn or worked by a horse or by horse-power" (*horse-barge, horse-railroad*). (Unfortunately, one of these headings consists of this very vague "definition": *of, pertaining or relating to, or connected with a horse or horses;* the predictable result is a most heterogeneous sampling of CNs, which mixes *horse-kick* with *horse-dentist, horse-market* with *horse-muck,* and so forth.)

The listing of so many expressions without specific definitions can only be motivated by a desire to be as (reasonably) complete as possible in listing CNs formed with *horse* as prenominal modifier; such an approach, however, commits the lexicographer to including both incomplete and redundant information in the dictionary. To begin with, the standard dictionary adherence to alphabetical listings entails in this case that CN forms may be grouped only by a common prenominal modifier, while CN forms that share a head noun (e.g., *farm horse, workhorse, circus horse, racehorse*) will be scattered throughout the dictionary. Since, however, the purpose of the grouping is presumably to demonstrate the productive ways in which a given noun enters into "combination" forms, this demonstration will always be at least 50% incomplete since there will be no way for a user to assemble the data required to show the range of prenominal modifiers with which a given head noun may combine.[6]

---

[5] A qualifying comment is in order. My training and experience lie primarily in lexical theory and not at all in practical or applied lexicography; I must therefore apologize in advance for whatever unintentional distortions my ignorance leads me to make. Certain oversimplifications have been made here deliberately, however, to suit present purposes. For example, I have ignored the quite likely possibility that the redundancies which one strives to eliminate in a theoretical description might serve some eminently justifiable function in a published dictionary, which after all exists to serve many diverse objectives and audiences.

[6] Strictly speaking, of course, the incompleteness will always be immeasurably greater since both the creative and the recursive properties which characterize CN formation mean that there is **no** way in which listings such as these could ever be exhaustive.

                                    6 IMPLICATIONS OF THE THEORY

The second major drawback of this system is one of redundancy. For example, this section of the *horse* entry includes 2 CNs under "appositive," 31 CNs under "for a horse," 17 CNs under "carried, drawn, or worked by a horse," 13 Agent Nominalizations and 10 other Objective NOM CNs under "objective and objective genitive," and 19 CNs under the catchall heading of "of, pertaining or relating to, or connected with a horse or horses." All of these categories except the last, however, represent fully productive patterns for forming CNs using **any** nouns in the language; in fact, they correspond to the demonstrably productive classes analyzed in this work under BE, FOR, USE, Agent Nominalization, and Objective NOM CNs, respectively. The listing of these individual categories under *horse*, together with over a hundred illustrative CNs, serves much more of an encyclopaedic function (by telling us something about the range of objects and activities associated with horses by English-speaking peoples) than an essentially lexicographic function, since the fact that CNs beginning with *horse-* form these particular semantic groupings is predictable from the grammar of CNs in general, and not from any property of the noun *horse* in particular.

An approach that is superior on both theoretical **and** practical grounds is that which Zimmer (1964) has suggested in a slightly different context:

> Where one is dealing with a clearly productive morphological process, a simple statement of the semantic content of the process in question, which would enable one to interpret new formations, seems to be as much as can or should be expected of a dictionary (together, of course, with a list of attested forms that are semantically specialized or irregular). A listing of semantically transparent attested forms (which in any case is in practice bound to be incomplete) is hardly less futile than an attempt to count the drops in a pool during a rainstorm. Moreover, it has to some extent the effect of obscuring the fact that the process *is* synchronically productive [p. 32].

Although Zimmer's remarks were made in regard to a more specifically morphological process (i.e., the formation of derived adjectives by prefixation with *non-*) than the syntactic processes involved in CN formation, his criticisms are in fact applicable to both these types of word formation and, in particular, to such practices as the inclusion of 116 **undefined** CN forms within the lexical entry for *horse*.[7]

---

[7] This figure does not include 119 additional CN forms listed with at least brief definitions as "special combinations" within the entry for *horse*, such as *horse-furniture*, *horse-nightcap*, *horse worm*, and *horse-violet*, nor the 74 CNs formed in *horse-* that are given full entries of their own; these categories presumably represent CNs which are either slightly or wholly idiosyncratic and hence require more detailed information. Along with Zimmer, I do not question the need to list such semantically specialized forms; nonetheless, the lexicographer (like the linguist) must take pains to apply as clear and consistent criteria as possible in distinguishing forms with predictable senses from those which are truly idiosyncratic.

The listing of CN forms that are presumably not idiosyncratic must therefore be recognized not only as an inherently interminable task, but also as one which is linguistically misleading, as Zimmer's final sentence properly indicates; in fact, the peculiar compromise reached by the *OED* in according a place to these CNs while at the same time denying them definitions amounts to an admission that they simply do not belong in a dictionary.

The exclusion from a dictionary of perfectly regular CN forms may be justified not only on the grounds of redundancy but on related grounds of inconsistency. For example, why should *horse* have over 300 related CN forms in (or near) its entry, while *camel* has a mere 30? Why is it that such forms as *horse-feed, -body, -dropping, -kick, -market, -side, -supply, -rug, -breeder, -dealer, -stealer,* and many more have been provided while the comparable *camel-feed, -body, -dropping, -kick,* and so forth have not? Surely there have been some English speakers who have had occasion to use the latter set of terms in certain parts of the once mighty British Empire.

The moral to this story is obviously that the disproportionate listings in the *OED* for these two sets of CNs tell us only about the disproportionate roles played by the horse and the camel in the history of English-speaking peoples, but nothing at all about the **grammatical** possibilities associated with the noun *horse* for entering into CN formation as opposed to the comparable possibilities for the noun *camel*. Although it is true that the content of any dictionary will invariably (and, in many cases, appropriately) reflect the cultural and historical background of the people(s) using the language in question (i.e., with information that is basically "encyclopaedic" rather than purely "lexical" in nature), this is certainly not the primary function of a lexicon in either the theoretical or applied sense of the word. In this particular case, it is clear that although the quantitatively different listings under *horse* and *camel* do correspond to certain "encyclopaedic" facts, they actually distort the linguistic fact that these two nouns have identical, rather than distinct, grammatical potential for appearing as prenominal modifiers (or head nouns) within CNs.

To summarize, we have seen that listings of this sort tend to add incomplete, redundant, and inevitably inconsistent information to the content of a dictionary. We may therefore conclude that the lexicographic practices typified here by examples from the *OED* (which we may assume is not markedly less sophisticated than any other modern English dictionary) in regard to listing semantically unspecialized CNs are without linguistic justification.

The practice of published dictionaries with respect to entries for nominal adjectives is even less satisfactory, since these listings are not

                          6 IMPLICATIONS OF THE THEORY

only inconsistent, incomplete, and redundant, but are often incorrect and frustrating as well. Most dictionaries have a "defining formula" for nominal adjectives which comprises such vague and unhelpful phrases as *of, belonging to, or of the nature of x* or *characteristic of or pertaining to x* (where *x* is the noun which the adjective replaces). For example, the OED defines the nominal adjective sense of *interjectional* as 'of, belonging to, or of the nature of an interjection in language' and that of *malarial* as 'belonging to, or of the nature of, malaria'; similarly, the *American Heritage Dictionary* (1969) defines *avian* as 'of, pertaining to, or characteristic of birds' and *urban* as 'pertaining to, located in, or constituting a city; characteristic of the city or city life'.

One major problem with these definitions is that they do not define; instead, they merely trade on that aspect of our linguistic competence that tells us what are possible (as well as what are plausible) readings for specific CNs. For example, the definition of *urban* just given does not in any way suggest that to interpret the CNs *urban problems*, *urban renewal*, and *urban lobbying* as, let us say, 'problems constituting a city', 'renewal characteristic of cities', and 'lobbying pertaining to cities' is any less plausible than interpreting them as 'problems that cities have', 'renewal in or of a city', and 'lobbying done by cities', respectively. The fact that native speakers regularly use and understand these CNs in the latter rather than the former senses means that the dictionary has **not** provided definitions that accurately reflect the realities of native speaker competence. The definitions they do supply are thus just as fallacious as the claims made by some linguists that the CN configuration means nothing more than that 'N_2 is related in some way to N_1'. (The untenability of this position was demonstrated earlier in Section 4.1.3 in connection with the RDP ABOUT.)

Another way in which these definitions fail to define follows from the fact that the inclusion of the genitive *of* in entries for nominal adjectives (e.g., 'of birds' or 'of electricity' within the entries for *avian* and *electric*) provides little more information than an observation that forms in which these adjectives appear are often paraphrasable by forms with corresponding genitive phrases (cf. *avian selection* and *selection of birds, electric current* and *current of electricity*); however, this does not bring us much closer to the meaning of these adjectives since the genitival forms typically display almost as much ambiguity as the CN forms with which they may alternate.

Yet even if we consider those aspects of dictionary entries that go beyond a simple *of* paraphrase (as indeed most do), we are still obliged to conclude that definitions of nominal adjectives like the ones cited here are unsatisfactory on the grounds of incompleteness and of redundancy.

These definitions are incomplete in that they never reflect the full range of meanings that are expressible in CN forms containing these adjectives; **if** one wishes to specify that *urban* can mean 'constituting a city or town' (as in *urban areas* or *urban locations*), then one ought also to specify that *urban* can mean 'possessed by a city or town' (as in *urban prerogatives*), 'done by a city or town' (as in *urban lobbying*), 'done to a city or town' (*urban renewal*), and so forth. Note, however, that all these senses are predictable from the grammatical analysis provided here for CN formation and thus need not be included, either partially or entirely, in the definition of every nominal adjective; such inclusion would make, and in actual practice does make, these definitions not only incomplete but also enormously redundant.

One additional problem that arises in the treatment of nominal adjectives in practical dictionaries is that in the vast majority of cases, no explicit or systematic connections are drawn between nouns and their nominal adjectives. Since the only organizational principle that comes into play in these dictionaries is that of alphabetical order, it is purely accidental that nominal adjectives like *malarial* and *hemispheric* end up immediately following their underlying nouns, whereas the adjectives *electric* and *harmonic* precede their corresponding nouns. Worse yet, nouns that have no morphologically related adjectives but rather suppletive adjectival equivalents (as in *heart/cardiac*, *spring/vernal*, *bird/avian*, *city/urban*, and many more) may be separated from their nominal adjectives by hundreds of pages. What is frustrating about this policy (or rather, nonpolicy) is that there is simply no direct way of finding out from these dictionaries what the adjectival substitute for a noun is, or even whether such an equivalent exists. This must certainly be annoying to a nonnative speaker, but similar frustration awaits even the native speaker who wishes to know, for example, whether there **is** a nominal adjective available that might replace a given noun, if only in technical language. Thus, I happen to know that *buccal* may replace *cheek*, and I accidentally discovered that *lacustrine* exists as an adjectival equivalent of *lake*, but I simply have no direct way of ascertaining from my dictionary whether nominal adjectives exist to replace such nouns as *chicken*, *harbor*, or *grass*.

The only exception to this pattern of silence may be found in dictionaries somewhat shorter than the *OED* where, presumably to save space, some nominal adjectives are simply listed without definitions at the end of the entry for their underlying nouns. In *The American Heritage Dictionary*, for example, adjectives like *interjectional* and *malarial* which are given definitions of their own in the *OED* are simply cited (with no comment other than a notation of *adj.*) at the end of the entries for *interjection* and *malaria*, respectively; similar treatment is accorded *hemispheric, mausolean, hemorrhoidal, meiotic, cybernetic,* and many others. Although this practice conforms much more closely to the linguistic facts of life in this area, this commendable conciseness unfortunately appears to be reserved for adjectives which are (*a*) relatively rare, and (*b*) derived by simple affixation from the basic noun. This policy therefore excludes the very many nominal adjectives of moderate to high frequency, as well as those formed by morphological suppletion.

Without dwelling any further on the problems faced by both the compilers and the users of printed dictionaries, let us now consider what positive recommendations may be made regarding the content and organization of a theoretical lexicon. Again, I will divide my recommendations into those pertaining to CN forms viewed as units, and those pertaining to nominal adjectives specifically.

With regard to listing CNs, it seems clear that the only CNs which should be entered in a lexicon are those whose meanings do not fall entirely within the range predicted by the general principles of our grammar of CNs; whether this semantic idiosyncrasy is extreme (as in such idioms as *duck soup, horsefeathers,* or perhaps *honeymoon*) or whether it is localized in just one component of the CN (as in *polka dot, monkey wrench,* or *cottage cheese*), the entire form must be listed in order to provide the unpredictable elements of its meaning.

These listings should **not,** however, be provided if their only purpose is to specify just which of the 14 theoretically possible readings of a given CN form is (or are) actually in common use; any attempt to do so for nonidiosyncratic CNs must end in incomplete, inadequate, and misleading information. For example, to include in a lexicon the specification that *doghouse* is more commonly used to mean 'house FOR a dog' than 'house FROM a dog', or that *apple cake* usually means 'cake which HAS apples' and not 'cake which MAKES apples' is to provide information for innumerable single entries that is predictable on the basis of very broad (and to some extent universal) semantic principles. Such an approach is to be avoided on several grounds: It would burden the lexicon with an enormous amount of redundant description, it would treat as idiosyncratic that which is not, and it would deny the semantic flexibility (i.e., potential ambiguity) that surface CN forms actually have. (Moreover, the listing of *doghouse* and not *lion house,* or *apple cake* and not *carrot cake,* leads to precisely the same inconsistencies as those brought out in the *horse* and *camel* discussion earlier.) Instead, the semantic principles from which these specific readings are largely, if not infallibly, predictable should be stated once in the grammar, and treated as part of the grammatical resources on which we draw when using CNs in actual discourse.

As for the proper treatment of nominal adjectives, every noun should have as part of its entry a specification of the nominal adjective (if any) which may replace it in CN forms; in addition, every nominal adjective will be given its own entry (which, in practical dictionaries, could also accommodate such nondefinitional information as phonology, etymology, or perhaps marked stylistic level) but with no **semantic** information provided beyond the statement that this adjective is equivalent to such and such a noun.[8] In this way, all possible types of noun–adjective pairs will be explicitly associated by cross-referenced entries: morphologically related pairs such as *malaria/malarial*, suppletive forms such as *city/urban*, and "defective" pairs where no nominal adjective exists at all, such as (presumably) *grass/∅* or *parakeet /∅*. (The possibility also exists for pairs to be defective in the opposite sense, namely, where an underlying noun surfaces only in nominal adjective form; one good candidate is *∅/civil* as in *civil engineering* or *civil service*, although the idiomaticity of these expressions somewhat weakens the case.)

This approach has the advantage of avoiding most of the difficulties that we have seen are entailed by the kinds of listings now found regularly in published dictionaries. In addition, two other advantages accrue from treating these adjectives basically as noun substitutes, and thus as a lexical category quite different from that of regular predicating adjectives. In the first place, this treatment permits us to predict that these adjectives are syntactically nonpredicating (a fact that is nowhere mentioned or even implied in the current dictionary practice of listing together both nonpredicating and predicating senses of adjectives such as *musical* or *electric*). Second, the distinct treatments permit us to predict that nominal adjectives must be negated by *non-* rather than by other negative affixes such as *in-* and *un-*, since nouns are also typically negated by *non-* (cf. *nonstudent, nondefinition, nonlinguist,* or *nonaggression*). We thus can disambiguate such expressions as *legal organizations* or *diplomatic historians* by comparing the two possible negations; the negation of the adjectives in the CN forms gives us *non-legal organizations* and *non-diplomatic historians*, while the negation of the predicating adjectives in the NPs that are not CNs produces *illegal organizations* and *undiplomatic historians*.[9]

One final suggestion for listing nominal adjectives is a more tentative one, namely, that the stylistic level achieved by the substitution of the

---

[8] At least the first of these suggestions is made explicitly by Apresyan, Mel'čuk and Žolkovsky (1970:10), as part of their lengthy proposal for a radical restructuring of unilingual dictionaries on a principled basis.

[9] This contrast in adjectival negation has been noted by Zimmer (1964:33–34) for English, and by Bartning (1976:80ff.) for French.

 6 IMPLICATIONS OF THE THEORY

nominal adjective for the noun be specified, at least in those cases where there is a marked difference between the two. For example, the use of *mind* and *mental*, *race* and *racial*, *society* and *social* entails little if any distinction in stylistic level, since all these forms belong to common, everyday, nontechnical discourse; in contrast, the substitution of *buccal* for *cheek*, *fluvial* for *river*, or *equine* for *horse* entails an extreme shift in style from the most basic level to a highly technical one. For a lexicon to state that a given noun and nominal adjective are equivalent when the contexts in which they are appropriately used are clearly quite disparate is to omit a description of one important part of the linguistic competence that we all share. This suggestion is offered only tentatively, however, since I have no idea as to how (or how many) stylistic levels might be usefully differentiated, nor do I wish to enter into the complex issue of what other kinds of contextual information ought to be included in an optimal lexical entry. Still, this notion offers some promise as a topic for further exploration.

6.3 Syntax, Semantics, and Pragmatics

The purpose of this study has been to explore the syntactic and semantic properties of complex nominals, and to propose transformational derivations (within generative semantic theory) which would make explicit the range of possible semantic structures that can be productively and predictably associated with a given surface CN. In the course of this exploration, it has become clear that a complete description of the role of complex nominals in natural language must include not only the kind of syntactic and semantic facts that formal derivations can account for, but also a description of the broader semantic and pragmatic principles that influence the ways in which both speakers and hearers manipulate the formal regularities of CNs in actual discourse. Although this latter aspect of the grammar of CNs lies outside the scope of this study, its indisputable relevance to our understanding of CNs as well as its intrinsic interest suggest that a brief discussion of the major issues may be appropriately included here.[10]

[10] My discussion in this section has been influenced significantly by the studies of compounding done by Zimmer (1971, 1972) and, more recently, by Downing (1975, 1977). Both these authors have focused primarily on contextual and pragmatic aspects of the compounding process, especially with regard to the coining of new CN forms; their work thus constitutes a valuable complement to the syntactic and semantic study of CNs represented by my own work. I have profited in particular from Downing's insightful analysis of the **function** of CNs in natural language (in which she builds upon and extends some of Zimmer's earlier proposals) and from explicit criticisms she has made (Downing 1975, 1977, personal communication) in regard to some of my earlier claims (in Levi 1973, 1974).

The syntactic derivations of CNs that have been provided in the preceding chapters permit us to associate a specific number of semantic structures with every surface CN; however, while these semantic structures are highly predictable in form and content, they are frequently also extremely vague. The reader will recall, for example, the discussions in Section 4.1 on the very vague predicates HAVE, FOR, and IN, and the justification given there for excluding more specific information from the underlying structures of at least nonidiosyncratic CNs. When familiar CN forms (e.g., *doghouse, garbage man, paper doll*) are used in discourse, the level of generality represented in the underlying structures I have proposed presents few problems for either speaker or hearer since encyclopaedic knowledge of the normal referent of that form requires no complex mental steps either to derive the form or to interpret it; it seems reasonable to assume that such common forms are both produced and interpreted with the same ease as other **lexical** items which, to be sure, have their semantic complexities but which are not so obviously the product of a regular syntactic derivation.

On the other hand, when a speaker creatively coins a new CN form and/or when a hearer encounters an unfamiliar one, additional steps must be performed by both speaker and hearer in order for the former's creativity (or erudition) not to interfere with successful communication. It is here that independently motivated, and more general, semantic and pragmatic principles must be brought into play by the speech act participants; only in this way will they be able to move beyond the level of generality represented in semantic structure (or, more precisely, in the set of possible semantic structures that they may associate with the surface form) to identify the specific, inevitably complex referent that is named by the new CN.

Some suggestions concerning the ways in which semantic and pragmatic considerations are used by listeners to figure out plausible referents for unfamiliar CNs were given earlier in Section 4.4. (As the reader may recall, the examples provided there illustrated both general semantic principles, such as those that help us determine that a locative reading is more plausible for CNs like *marginal note* and *sea turtle* than for *musical clock* or *cough syrup*, and pragmatic strategies through which we apply our extralinguistic knowledge of contexts and culture in order to disambiguate CNs such as *musical clock* and *vehicular vermin*.) The basic function of these principles, when used in conjunction with the syntactic and semantic constraints on CN formation, seems to be one of helping the listener to select from the grammatically **possible** semantic structures the one reading that is contextually most **plausible,** and then (or, more probably, simultaneously) to figure out what real world object could be appropriately named by such a form. The speaker's task, on the other hand, is closely related to that of the listener, in that the speaker must choose a CN form in such a manner as to permit relatively prompt and accurate identification by the listener of the intended referent.

Any attempt to figure out just how speakers and listeners go about their various tasks must take into account the fact that we bring many different kinds of knowledge about CNs to our daily conversations. To begin with, we have knowledge of the general, recursive rules by means of which we associate a given surface CN with a given set of semantic structures; these rules, which have both universal and language-specific aspects, correspond to the transformations making up the derivations of Chapters 4 and 5. Let us categorize these rules, at least for present purposes, as a part of our "strictly grammatical" knowledge. (This is the only kind of knowledge for which formalized statements have been attempted in this book.)

In addition, we know which particular semantic structure is **most commonly** and/or **most plausibly** associated with certain surface forms; this could be called our knowledge of the "common" or "institutionalized" reading of a given CN. (The term "institutionalized" is here adapted from Matthews 1974:193.) We must note that this knowledge is a rather flexible and even ephemeral sort since it is sensitive to changes in the extralinguistic environment in a way that our "strictly grammatical" knowledge of CNs is not; for example, the meanings that we now associate with such CNs as *rock music, nose cone, nose job, leisure suit, liver scan, copy machine, iron curtain,* and even *surface structure* are very much a function of developments in the real world in just the last few decades, and thus could shift to other readings (or even fall into near total disuse) as a function of future developments of a comparable extralinguistic nature.

Other kinds of knowledge that affect our usage of CNs include straightforward **lexical** knowledge of the meaning of idiosyncratic and metaphorical CN forms such as *polka dot, honeymoon,* and *stool pigeon,* and what might be called **stylistic** knowledge which helps us select appropriate contexts in which to use variant or periphrastic forms of CNs (as illustrated earlier by the discussion in Section 6.1).

It seems reasonable to establish as an objective that all these kinds of knowledge be recognized in a full account of the ways in which (and the levels at which) we use CNs to communicate. What is not yet clear is how best to represent those kinds of knowledge that are not "strictly grammatical" in the common sense of the term (especially the basically ephemeral facts pertaining to "institutionalized" readings of CNs, and the highly context-sensitive variables which enter into our stylistic judgments), or how these different but related kinds of knowledge may best be integrated within a single description.

Turning now to a consideration of the **semantic** principles that affect our selection and interpretation of CNs, we must note to begin with that the combinations of nouns that occur within a single CN are not entirely random but, rather, appear to follow certain highly favored semantic patterns. For example, an examination of attested CN forms in English reveals that the combinations shown in (6.4) are among the most frequent patterns, indicating that the ways in which English speakers choose to **name** particular objects or events in their world often (though not invariably) reflect a small set of organizing principles easily expressible in terms of semantic classes.[11]

(6.4)

Head N	Modifier	Examples
living thing	habitat	*field mouse, water lily, desert nomad*
artifact	function	*drill press, culinary utensils, induction coil*
	power source	*steam iron, electric drill, hand brake*
activity	time	*morning lectures, night flight, spring training*
	place	*home instruction, urban riots, office work*
	subject	*papal appeal, student demands, royal orders*
	object	*car repair, word formation, child abuse*
	instrument	*shock therapy, solar heating, smoke signals*
people	sex/age	*boy genius, woman lawyer, girlfriend*
	home (= habitat)	*country cousins, mountain tribes, city folk*
	occupation	*lawyer friends, student renters, clerical enemy*

Although these patterns were derived from English data only, it is highly likely that the principles by which living things are named by their habitats, artifacts by their purposes, and activities by the time and place of their occurrence represent not an English-specific phenomenon, but rather a set of universal semantic and cognitive principles according to which human beings everywhere structure their world.[12] I would therefore predict with considerable confidence that essentially similar patterns of semantically definable noun combinations will be observed in comparable collections of CN forms in other languages; this represents, in fact, a very promising line of crosslinguistic investigation. Still, we must keep in

---

[11] Clancy (1975) and Downing (1975, 1977) provide fuller discussions of this idea, based on their respective psycholinguistic studies in this area.

[12] The affirmation of universal semantic principles of CN formation in no way precludes the possibility of there being language- or perhaps culture-specific principles in this area (a topic to be discussed in the next section).

mind that the prediction (or verification) of which CN patterns are most **likely** to occur in a given language is a separate issue from the question of which CN forms are **possible** in a given language; only the latter question has been the subject of inquiry here, while the former task is one that still awaits future research.

In addition to applying semantic principles like those exemplified by the patterns shown in (6.4), native speakers exploit a number of **pragmatic** principles in producing and understanding CNs; the latter are not directly tied to the component parts of specific CNs, but rather represent principles and strategies of considerably broader scope whose primary purpose is to facilitate efficient and successful communication.

Let us first consider these principles from the viewpoint of a speaker who decides to use a novel CN form (or a familiar CN form to denote an unfamiliar referent). Let us also assume, along with Zimmer (1971:C15–C16) and Downing (1975:40ff.), that such acts of linguistic creativity occur when a speaker wishes to denote an object by means of a concise name rather than by a more explicit but lengthier descriptive phrase (e.g., *my volleyball friends* rather than *friends whom I know from playing volleyball*), but no well-established name (in CN form or otherwise) exists for the object in question. What pragmatic principles must the speaker observe in order that the two nouns newly adjoined in a CN form will permit the listener to identify the intended referent correctly?

To begin with, the speaker must observe the convention that CNs are normally endocentric; the head noun chosen must therefore describe a superset from which the modifier will carve out a more narrowly defined subset. The speaker is then faced with the more difficult task of selecting a prenominal modifier that will enable the listener to pick out just the right referent (and, not incidentally, to understand the speaker's reason for choosing just that name and no other). One of the most important pragmatic principles is that the prenominal modifier must denote some truly **distinctive** feature which will isolate the appropriate subset from all others; two corollaries of this principle are that this feature must not express either negative relations or redundant information. For example, it would be redundant to use a CN such as *water lake* or *heat cooker*, in view of our knowledge (or belief) that all lakes have water and all "cookers" cook by means of heat; these features are therefore predictable from the head noun. In contrast, the familiar CNs *salt lake* and *pressure cooker* are functional, CN names because lakes that have (unusual amounts of) salt and "cookers" that use not only heat but also air pressure to cook their contents are distinctive enough to constitute special subsets.

On the other hand, we could not effectively use *salt lake* to mean 'lake which has no salt in it' or *pressure cooker* to mean 'cooker which uses any means except pressure to cook' since naming something by a characteristic that it does **not** have is so obtuse a strategy as to be wholly dysfunctional. As Downing (1975) notes:

> The reason for the unacceptability of compounds based on non-relationships is intuitively apparent. Classifying a given individual by virtue of its membership in the set of individuals with which a given entity or quantity is not associated does not narrow down the class of referents to any useful degree, since there are typically an infinite variety of individuals which do not have a particular quality, and an infinite variety of qualities which a given individual does not have. Thus a categorization based on reference to an existing relationship ordinarily carries more information than reference to a non-existing one [p. 58].

What is interesting about these remarks is that they represent one of the ways in which Downing's careful attention to the **function** of CNs illuminates an otherwise unexplained aspect of the syntactic–semantic study presented here, namely, the absence of NEG from the RDP set. Although this set includes many other semantically basic relationships, the reader may recall that NEG is actually conspicuous by its absence. We can now understand the exclusion of NEG from the RDP set, however, in view of the dysfunctional effect on communication that its inclusion would create.

It also follows from this pragmatic principle that the modifier chosen for the CN name need not encode **all** information about the intended referent but only the most salient characteristic needed to identify it. For example, although we know that what we call *gingerbread* must have such customary ingredients as sugar, shortening, and flour, such information would not be effective in distinguishing that food denoted by *gingerbread* from various other baked goods; the selection of the modifier noun *ginger*, however, does serve this purpose since the inclusion of a noticeable amount of ginger in a bread or cake is unusual (for most American palates, at least).

In addition to this primary pragmatic principle, we apparently also observe the principle that naming an object in terms of some regular, habitual, or permanent association or relationship will be more helpful to the listener than selecting a CN based on some totally ephemeral juxtaposition of features. Thus, if I tell you that I have seen an *aquatic bird* this morning, but I mean a chicken that had been thrown into Lake Michigan, I would be violating this principle since you would (justifiably) have assumed I meant a bird that is habitually rather than fortuitously around water, such as a duck or a seagull. Somewhat similarly, the fact that I am occasionally obliged to teach the rudiments of English grammar to students in a general linguistics course does not make me a *grammar teacher* in the normal sense of the term.

These two principles of distinctiveness and permanence may be taken as illustrative of the kinds of pragmatic principles which guide speakers in determining just which of the numerous possible CN forms will be the most suitable for purposes of effective communication as well as creative expression; there are undoubtedly others that may occur to the reader, although no additional suggestions will be included here.[13] Instead, I will conclude this section with three final observations.

The first of these is that the task of the listener who hears an unfamiliar CN essentially consists of retracing the path followed by the speaker in associating a new CN with a particular referent. Despite the ambiguity (often compounded by vagueness) which characterizes CN forms, this task is made possible by the fact that both speaker and listener know "the rules of the CN game," that is, they share a knowledge of the semantic and pragmatic principles associated with CN formation. The listener then **assumes** that these principles have been invoked (e.g., that the CN is an endocentric one, that the modifier was chosen to denote a positive rather than negative characterization of the relevant subset, and that the relationship between the two nouns is a nonephemeral one) and goes on from there to pick out the most likely referent for the given form and context. Although this task is not one for which immediate success is guaranteed, the principles just discussed are remarkably effective for this purpose.

Note also that the problem of the listener has been described here in terms of correctly identifying the **referent** that the speaker had in mind; this is inherently a different problem than identifying the underlying structure from which the speaker presumably derived that CN. The pragmatic reality of human discourse requires only that the former be achieved, whereas accomplishment of the latter is sometimes more fortuitous than essential. For example, if a speaker imagines *musical talent* to have a meaning roughly like 'talent in [the field of] music' and a listener interprets the same CN as 'talent for [performing or creating] music', the fact that the former could be said to derive the CN by IN Deletion (from *talent in music*) while the latter would derive it by FOR Deletion (from *talent for music*) is unlikely to interfere with successful communication. Similarly, if a speaker thinks of *muscle weakness* as 'weakness in a muscle', it does not matter in this case if the addressee interprets this CN as 'weakness that a muscle has' or even as a nominalized adjective with a meaning like 'the fact that the muscle is weak' or 'the state of a muscle's being weak'. This kind of hidden ambiguity in most cases poses no

[13] It is quite possible that this brief list of pragmatic principles might be appropriately extended by adapting some of the Cooperative Principles proposed by Grice (1975) to the specific area of communication via CNs; the maxims included under Quantity, Relation, and (to a lesser extent) Manner appear promising in this regard.

obstacle to mutual comprehensibility, although it cannot be denied that this is a potential source of confusion. (For example, if a speaker used *cat spray* as a Subjective Patient Nominalization equivalent to 'that which cats spray' to refer to that malodorous product with which male cats leave their marks, a listener who interpreted the CN as 'spray for cats' along the lines of *hair spray* or *insect spray* would be seriously off target in understanding the speaker's intent; the lack of agreement between speaker and hearer might eventually become apparent, or it might persist if subsequent remarks gave no hint to either speech act participant that they had had different referents in mind.)

This brings us to the last of these concluding remarks, which is simply a reminder that all of these principles—both semantic and pragmatic—jointly serve the function of helping us reduce the multiple ambiguity and inherent vagueness of these CN forms in ways that a purely grammatical description does not, and should not, do. Although these principles do not **preclude** the possibility of misunderstandings and misinterpretations of CNs in discourse, they nonetheless reduce it sufficiently so that the expressive concision and stylistic flexibility that the unique CN construction has to offer can be systematically and creatively exploited in natural language.

6.4 Complex Nominals and Universal Grammar

The subject of linguistic universals was first raised in Section 4.5, where I pointed out that the semantic primitiveness of the RDP relations, combined with their unusual behavior in areas of the grammar unrelated to CN formation, strongly suggests that this group of predicates not only is significant for the formation of CNs in English but is also distinctive on a much wider and quite probably universal scale. In the present section, a somewhat fuller consideration of the implications of this study for universal grammar is provided, and an attempt is made to distinguish between those aspects of the grammar of CNs that are inherently language-specific and those that are most likely to be reflections of linguistic universals in this area.

Although all proposals in the area of linguistic universals which precede extensive and systematic crosslinguistic studies must be regarded as highly tentative, I have based the present discussion not only on my own analyses of English and Hebrew CNs (the latter in Levi 1976), but also on a variety of relevant studies done by others on such diverse languages as Afrikaans, Chinese, French, Japanese, Sanskrit, and Turkish.[14] Despite

[14] Specifically, I have profited from reading Barbaud (1971) and Bartning (1976) on CNs

the fact that this particular group of languages is still only a small sampling, I believe that it affords enough variety to justify the elaboration of a number of preliminary hypotheses.

Evidence from many sources indicates that the process of CN formation in any natural language is governed in part by a network of syntactic, semantic, and pragmatic universals and in part by a system of language-specific modifications and preferences. Let us consider first the most likely candidates for **universal** principles.

To begin with, we have good reason to hypothesize that the set of predicates that can be deleted in the process of forming CNs in natural language is a **universal** one; this implies that in every language that permits CN constructions, each and every member of this set will be found to underlie an appreciable number of both "institutionalized" and creatively formed CNs (although the relative frequencies are likely to vary from predicate to predicate and, perhaps, from language to language). Let us call this hypothesis the Strong Universal Hypothesis (SUH).

A weaker version of this hypothesis would be one that claimed that a universal set of deletable predicates exists but that CN formation in a particular language may involve only a subset of the universal set, such that any one of the member predicates might be excluded by a language from playing a productive part in the derivation of its CN forms.

There are, however, several reasons to advance the stronger version of the hypothesis here. First, as an intrinsically more powerful claim, it makes more interesting and more challenging predictions (and, concomitantly, is more easily falsifiable). Second, the predictions made by the weaker version (at least as stated here) could be satisfied not only by a world in which the set of RDPs differed from language to language by just 1 or 2 members out of some 9 or 10 (a situation that is quite conceivable, given present knowledge),[15] but also by a world in which the number of

---

in French, Botha (1968) on nominal compounds in Afrikaans, Chryst (1976) on the N–*no*–N construction in Japanese, Dede (1977) on Turkish nominal compounds, and the section on compounds in Whitney's classic study of Sanskrit (1879); in addition, the discussion in Downing 1975 on Li's analysis of Chinese compounds (Li 1971), and conversations and correspondence with Müşerref Dede of the University of Michigan on compounding processes in Turkish, have helped me to broaden my perspective.

[15] One piece of evidence that suggests interlanguage differences in RDP membership is the existence in Sanskrit of *dvandva* or "copulative" compounds (Whitney 1879:428), in which two or more nouns which are underlyingly conjoined appear on the surface in a series without overt expression of the underlying conjunction(s); in the present framework, this would be analyzed as CN formation by AND Deletion, a possibility that does not show up as such in English. (Note that coordinate forms like *secretary–treasurer* and *speaker–listener* were excluded from the present analysis on the grounds of their exocentricity. In addition,

RDPs in different languages could vary anywhere from, say, 100 or more down to the 9 which have been proposed here for English, all the way down to just 1. Since the second of these two possible worlds seems remote indeed from linguistic reality, while the first seems much closer in spirit (or structure) to what the SUH would predict, we can make more progress in this area by advancing the stronger claim and then seeing how well or poorly it performs in the field.

The Strong Universal Hypothesis advanced here does **not** include the claim that the enumeration and description of specific RDPs in Chapter 4 necessarily constitute an optimal analysis of the relevant data; indeed, quite legitimate questions may be raised concerning such topics as the unity and validity of the predicate MAKE, the unusually restricted productivity of both ABOUT and FROM, or even the possibility that the predicates CAUSE, MAKE, and FROM are all related at a deeper semantic level in a way that is not captured in the present analysis. There is nonetheless much evidence to support the SUH and, in any case, its validity must be examined as an issue that is independent of whatever degree of accuracy and insightfulness any individual investigator manages to achieve.

Two other universals of an essentially **syntactic** nature that might be proposed are (*a*) that CNs are universally formed by the adjunction of just two nouns at a time, producing the exclusively binary branching patterns illustrated earlier in (4.51); and (*b*) that nominalization processes **tend** to produce structures that are isomorphic in form with CNs produced by predicate deletion. (This is not proposed as an absolute universal for the simple reason that I have seen nothing in the literature, beyond Lees 1960 and my own studies of English and Hebrew CNs, where the outputs of "compounding" and nominalization processes are explicitly compared; the silence of scholarship thus impels me to caution.)

The question of **semantic** principles used to construct new CNs and to disambiguate unfamiliar ones was raised in some detail in the preceding section; the principles suggested there, according to which nouns of

---

English forms which superficially resemble *dvandva* compounds, such as the modifiers in *love–hate relationship*, *boy–girl arrangement*, or *town–gown tensions* are not relevant here since they never appear as independent nominals but only as prenominal modifiers; they therefore could not be derived by the same processes that form CNs, since every preposed modifier in a CN must at some point in the derivation have constituted a single, independent, and grammatically well-formed noun.) Examples of the *dvandva* forms include *satyānrté* 'truth and falsehood', *rksāmé* 'verse and chant', *candrādityāu* 'moon and sun', and the longer *devagandharvamānusoragaraksasās* 'gods and Gandharvas and men and serpents and demons' (Whitney 1879:425ff.). Note, however, that the possibility of conjoining three or more nouns in this *dvandva* construction suggests a fundamentally different process from the binary adjunction characteristic of all CNs analyzed here.

   6 IMPLICATIONS OF THE THEORY

certain semantic classes are more likely to be joined by some RDPs than by some others, are excellent candidates for semantic universals. In addition, inspired by parallels already found between English and Hebrew CNs (see Levi 1976:15), we might hypothesize that the relative productivity of the various derivational types of CNs would show substantial consistency across languages, although it is impossible to guess at this stage just how uniform such distributions actually are. (Two important problems to resolve in this area are how to measure "relative productivity," and the extent to which cultural rather than linguistic differences significantly affect the distribution.)

In the area of **pragmatic** universals, we can hypothesize the following with considerable assurance: (*a*) that the naming function of CNs will be constant across languages; (*b*) that the use of permanent or habitual relationships rather than ephemeral or accidental ones to form CNs will predominate everywhere; and (*c*) that whatever disadvantages are created by the multiple ambiguities associated with CN forms will be outweighed for speakers of most languages by the expressive conciseness and functional utility which such forms typically provide. Finally, we can expect that the Gricean principles of conversational "perspicuity" (Grice 1975) will normally be observed by speakers throughout the world in producing and processing CNs in their respective languages.

In contrast to these universal patterns, the ways in which languages differ with respect to CN grammar and usage are likely to be of a more superficial nature. For example, we should expect to find crosslinguistic differences in the form, roles, and distribution of CNs and CN-like expressions. More specifically, some languages lack a Morph Adj rule for creating nominal adjectives from nouns in CNs, while those languages that have such a rule differ widely in how often it is applied; at the freer end of the spectrum we find French, which has a remarkably rich assortment of noun adjectivalization processes (cf. *présidentiel, laitier, chevalin, électrique, commercial, agricole, immobilier, européen, lunaire, manuel, terrestre, cardiaque, urbain*), whereas Hebrew (which comes from a language family favoring the N–N version of a CN) and Turkish would be ranged closer to the restricted end. In other cases, it is the N–N form of a CN that is disfavored or even prohibited as a result of language-specific constraints against the simple adjunction of two unmarked nouns without an accompanying grammatical morpheme inserted at one point or another; in fact, the typically ambiguous "genitive" marker (e.g., English *of*, French *de*, Modern Hebrew *šel*, Turkish *-i*, and Japanese *no*) is just such a morpheme, which some languages treat as obligatory everywhere and others as more of a stylistic option. In these and other cases where the language permits a variety of grammatically equivalent forms, we should also expect to find language-specific idiosyncrasies showing up in the stylistic values attached to each option, such as those discussed earlier in Section 6.1.

Another area in which languages can be expected to differ concerns the relation between CNs formed by nominalization and those formed by predicate deletion; it is quite likely that some languages would require minor morphological distinctions between the two types. Nonetheless, in view of the fact that the two processes typically produce surface forms of comparable compactness and concomitant usefulness, we would not be surprised to find most languages assigning a common target structure to the output of both predicate deletion and predicate nominalization rules. Needless to say, this expectation can be viewed only as a speculation until much more crosslinguistic work in this area has been carried out.

Other ways in which universal principles of CN formation might be modified from language to language include a number of possibilities raised earlier in the discussion, such as (*a*) minor differences in the set of RDPs involved in CN formation; (*b*) the association of slightly different syntactic phenomena with NOM CNs than with RDP-type CNs (as preliminary evidence has suggested for Hebrew); and (*c*) differences in productivity of the various CN types. And finally, languages may differ to some degree in the readings most commonly attributed to CNs made up of nouns of particular semantic classes. One clear example of this difference shows up in comparing English and Hebrew CNs used to denote a container and its contents; in Hebrew, CNs belonging to the same semantic class as the English equivalents *wineglass* and *jam jar* are regularly used with either a FOR or a HAVE₁ reading (e.g., as 'glass for wine' or 'glass with wine'), in contrast to English usage which tends to reserve the CN form for the FOR reading and to use a periphrastic genitive (*glass of wine, jar of jam*) for the HAVE₁ reading (Levi 1976:28–29). (Note that this sort of distinction does not involve questions of permissibility versus total unacceptability, but rather the relative **frequency** with which a language associates different derivational patterns with nominals of particular semantic classes; as a consequence, these distinctions are neither predicted nor precluded by my theory.)

One must always be wary in approaching the subject of linguistic universals in any area since one's own experience inevitably falls lamentably short of encompassing any appreciable range of language types and structures. On the other hand, to ignore the very suggestive evidence concerning universal principles of CN formation that the present study has uncovered would be to adopt an apparently prudent but regrettably shortsighted approach to the data in place of actively exploring what promises to be one of the most fruitful aspects of this entire enterprise.

                                   6 IMPLICATIONS OF THE THEORY

7

EXCEPTION CLASSES AND OTHER PROBLEMS

7.1 Classes of Possible Exceptions

In this section several different sets of data are presented, which appear to be unaccounted for by the derivations of Chapters 4 and 5; for several of these sets, I argue that the data are only apparent counterexamples, whose derivation is most accurately characterized by rules outside the area of CN formation. These sets of putative counterexamples include CNs formed by what have been called "suffix words" (Wentworth and Flexner 1975), CNs whose modifiers denote units of measure, and CNs whose adjectival modifiers must be derived from adverbs rather than nouns.[1] A simple list is also provided showing a residue of CN forms for which I have no explanation at all.

[1] One additional set of alleged counterexamples is exemplified by the many N–N expressions formed in recent years with *Watergate* as a prenominal modifier (e.g., *Watergate episode/tragedy/morality/judge/mentality/era/turmoil*). Although of some intrinsic interest, these forms are not directly relevant here since the use of a single noun such as *Watergate* to "label" a situation or event (whose reference is then construed very broadly, and in different ways by different speakers) is a phenomenon that is not limited to CN formation but rather occurs in a variety of linguistic constructions (cf. *the tragedy of Watergate, Washington after Watergate, the post-Watergate morality*, etc.). The analysis of these "label nouns" thus involves the issue of semantic extension (by metonymy) of such proper nouns in general, rather than their use in CN formation in particular.

In trying to decide which forms may be true counterexamples to a theory, one must first have clearly in mind the intended scope of the theory (i.e., the kinds of forms or phenomena that the theory is designed to account for, as well as those grammatical facts that are excluded a priori from its predictive field). In the present case, the scope of my theory was restricted (Section 1.3) in such a way as to exclude from consideration exocentric CNs, idiosyncratic or lexicalized CNs, and CNs whose modifiers are derived either from proper nouns or from underlying adverbs; these sets of CNs account for many, though not all, of the putative exceptions.

One must also be on the lookout for forms that share the target structure of the CNs analyzed here, but which can be shown to be derived from quite different underlying structures and so must undergo significantly different rules in order to emerge in the same surface guise. Explicit scope restrictions and nonparallel derivations thus represent the two most important guidelines to be used in determining the force of the alleged counterevidence to be analyzed here.

The first interesting set of potential counterexamples whose meanings (and hence, derivations) my theory does not predict is illustrated by the data in (7.1).

(7.1)   a. *coffee/leg/breast/pipe/baseball man*
           b. *cat/morning/tennis person*
           c. *volleyball/poetry/alfalfa/fitness/sitar freak (fiend, bug)*
           d. *vodka/shopping/studying/yogurt/sweater binge*
           e. *lunar/solar/fiscal/academic year*
           f. *dental/medical/psychiatric/gynecological appointment*
           g. *iron/bronze/industrial/space age*
           h. *salad/hatcheck/coffee/copy girl*
           i. *soap/diamond/brewery/textile heir(ess)*

Although all these forms have the superficial appearance of CNs, it is my claim that they actually represent a different type of N–N adjunction, formed not by the syntactic processes of predicate deletion or predicate nominalization but rather by a morphological process equivalent to suffixation; in this process, the aspects of meaning that are not overtly expressed (e.g., that *coffee man* means 'man who likes or prefers coffee' or that *soap heiress* means something like '[fem.] inheritor of large sums of money made from selling soap') are actually incorporated as part of the semantic structure of the head noun. Since the head noun thus functions precisely like such bound suffixes as *-phile* 'lover of' (cf. *Francophile, bibliophile*), *-phobia* 'fear of' (*hydrophobia, claustrophobia*), or *-iana* 'sayings and anecdotes about' or 'artifacts pertaining to' (*Shakespeariana,*

Johnsoniana), it may be appropriately dubbed a "suffix word."[2] Since a full explanation of the specific "suffix words" in (7.1) would take us too far afield, I will use just two examples to illustrate the general approach intended here. Specifically, we may note that the colloquial *freak* (or the equivalent *fiend* and *bug*) of (7.1c) is semantically comparable both to the more dignified *aficionado* and to the bound suffix *-phile* just cited, while the equally informal *binge* in (7.1d) carries a consistent meaning along the lines of 'exceptionally unrestrained indulgence in'.

Despite the rather impressive productivity observable for these formations, they must be analyzed as fundamentally different in source and derivation from normal CNs. This may be seen from the fact that the apparently covertly expressed aspects of meaning (such as those just mentioned) are in fact regularly "recoverable" from the head noun alone, a fact which has as a consequence that the head noun is virtually invariant. Thus, although *coffee man* may be used to mean 'man who likes coffee', more complex head nouns may not be substituted to express the same relationship; although *coffee girl* may be marginally acceptable, *coffee lady* and *coffee woman* sound unusable in this sense, and *coffee librarian* or *tea politician* strike me as simply unacceptable ways of saying 'librarian who likes coffee' or 'politician who likes tea'. Since, however, neither of these conditions characterizes any of the numerous CN types (i.e., it is never possible to recover the unexpressed aspects of meaning of a CN from the head noun alone, nor is there any derivational type of CN that virtually precludes variation in the head noun), we are justified in concluding that the data in (7.1) are not true counterexamples to the theory but rather represent a case of a shared target structure reached by means of quite distinct derivational paths.

A second and rather limited set of potential counterexamples consists of CN-like expressions whose prenominal element consists of a noun denoting a unit of measure (typically one of cost or length); these potential challenges to the theory appear in (7.2).

(7.2)   *penny candy, nickel cigars, yardstick, mile race*

There are several reasons for rejecting the expressions in (7.2) as true counterexamples to the present proposals. First of all, they constitute a severely restricted set which cannot be productively extended; thus, we do not speak of such objects as **dollar books, *inch seams, *ounce packages*, or **ton walruses*. Since, however, the meanings intended by these made-up CNs may be expressed in the phrases *one-dollar books, one-inch*

seams, one-ounce packages, or *one-ton walruses,* we may conclude that
the forms listed in (7.2) are actually idiosyncratically abbreviated versions
of the synonymous (if less common) expressions *one-penny candy, one-
nickel cigars, one-yard stick,* and *one-mile race.* A second reason for
treating these forms as non-CNs stems from the fact that they derive their
prenominal modifiers not from unaffixed generic nouns (as we have seen
to be true for CNs in general) but rather from nouns that have been
compounded with numerals (specifically, with the numeral *one*); they
thus must share a derivation with other expressions in which this compound modifier surfaces intact, as in *two-dollar cigars, eight-penny nails,
ten-mile race,* and *two-ton truck,* rather than with CNs themselves. The
third and last reason to reject these examples as counterevidence stems
from the fact that the most plausible way to change the theory so that it
would predict such forms would be to enlarge the set of RDPs to provide
suitable underlying predicates. However, since this would presumably
require the addition of both a COST and a MEASURE predicate (each of
which would generate a miserably small number of forms), and the
inexplicable exclusion of a comparable WEIGH predicate (since we apparently have no forms such as **pound cheese* or **ounce letter*), it seems
reasonable to conclude that the data in (7.2) constitute no set of systematic exceptions to the theory but only a very small (and irrelevant) collection of idiosyncratically abbreviated expressions.

Still another small set of examples whose meanings and derivations
are difficult to predict in the present theory may be explainable as instances of nonpredicating adjectives whose immediate derivational ancestors are adverbs, rather than nouns. Examples of this set, together with
related forms with overt adverbs, are shown in (7.3).

(7.3) a. *physical collapse* *They physically collapsed.*
     b. *sexual attack*   *She was attacked sexually.*
     c. *racial motivation* *The decision was racially motivated.*
     d. *central heating*   *It was centrally heated.*

The suggestion that these forms are derived not by the regular processes
of CN formation but rather by whatever process transforms adverbs in *-ly*
into prenominal nonpredicating adjectives is of course more an exercise
in passing the analytic buck than a detailed explanation; nonetheless, it
does seem plausible that these, and perhaps a few other, recalcitrant
examples are most appropriately derived from antecedent adverbs and so
do not constitute true exceptions. (For a more detailed discussion of
nonpredicating adjectives derived from antecedent adverbs, see Levi
1975, Chapters 7 and 8.)

Although I claim that the data in (7.1)–(7.3) are not true counterexamples to my theory, there remains nonetheless a small residue of

forms whose meanings and derivations are simply not predictable from the analysis of CNs given here; these forms are listed in (7.4). (Questioned material in parentheses indicates possible analyses which would fit within the framework of my theory.) In accordance with the scope restrictions alluded to at the beginning of this chapter, idiosyncratic forms showing no contemporary productivity at all (such as *cranberry morpheme*, *thalidomide parents*, and *tooth fairy*) have been excluded from this list as forms whose meanings no grammatical theory could be expected to predict. On the other hand, the fact that more than one prenominal modifier is in common usage for most of the examples in (7.4) suggests that the forms listed here do have at least a restricted productivity which English speakers apparently recognize but which the present theory does not yet reflect.

(7.4)   *milk/dairy/oil/publishing/education industry* (FOR?)
         meteorological/linguistic/topological/dialect map (Agent Nom?)
         snob/student/kiddie appeal
         student/executive/customer types
         land/air/water/oil rights
         race/religious riots
         racial/religious disorders
         dew/timber line
         violin/chess prodigy (IN?)
         chicken/sparrow hawk
         saturation temperature
         date line
         age groups

This list is certainly a small one. Its brevity is due in part to my having excluded on methodological grounds all forms that I judged to be idiosyncratic, and perhaps (in much smaller measure) to the randomness with which I have encountered and subsequently catalogued CN forms over the years. It is nonetheless crucial to observe that the number of nonidiosyncratic CN forms whose meanings and derivations my theory cannot account for is very small indeed in comparison to the thousands of CN forms that can be easily and regularly generated by the rules of Chapters 4 and 5.

7.2 Nonpredicating Adjectives in Predicate Position

The original impetus for my research came from the puzzling characteristics of those adjectives that I have called (**nominal**) **nonpredicating**

adjectives;[3] this term was chosen to emphasize the fact that these adjectives are regularly excluded from the predicate position so freely occupied by "normal" predicating adjectives. This fundamental difference in syntactic origin is reflected in the derivation of nominal adjectives provided in Chapters 4 and 5, where the only source of these adjectives is shown to be single nouns that occupy left-branch nodes in CN configurations. This contrasts sharply with the familiar transformational analysis of predicating adjectives, which are uniformly derived from a clause in which they occupy predicate position as independent (adjectival) constituents.

The validity of this otherwise well-motivated distinction is, however, apparently called into question by the undeniable fact that sentences like those in (7.5) are frequently heard (and easily interpreted) in normal English discourse.

(7.5)   a. *The process by which compounds are formed is transformational.*
         b. *Her infection turned out to be bacterial, not viral.*
         c. *His razor is electric.*
         d. *Question formation in Finnish is morphemic.*
         e. *The therapy David does is primarily musical.*
         f. *That interpretation is presidential, not judicial.*

Since I would certainly claim that the adjectives appearing in predicate position in (7.5) are tokens of the same adjectives that have been called nominal adjectives throughout, it is clear that an explanation for their somewhat unexpected appearance in these examples must be sought.

One argument that might be advanced to explain the data is that the use of these adjectives in predicate position represents an ad hoc syntactic generalization, reflecting an attempt by some speakers to regularize the syntax of surface adjectives. That is, since the syntax of "normal" (predicating) adjectives permits them to occur in both prenominal and predicate positions, and since adjectives like those in (7.5) share the same morphological characteristics of "normal" adjectives, it might be that speakers occasionally generalize the distributional patterns of predicating adjectives to nominal adjectives as well.

The problem in characterizing the data as a syntactic generalization is that such a characterization is at best only a description, not an explanation. Like the much invoked and much maligned process of analogy, the process implied by this use of the term "generalization" explains neither

---

[3] As noted earlier, the term *nonpredicating adjectives* properly includes both those derived from nouns (*nominal nonpred adjs*) and those derived from adverbs (*adverbial nonpred adjs*); since only the former are pertinent to the present work, the discussion to follow will use *nominal adjective* and *nonpredicating adjective* interchangeably.

     7 EXCEPTION CLASSES AND OTHER PROBLEMS

why the change took place, nor why other changes that the same process might have produced did not take place. (For example, it is still the case that we cannot say *very governmental property* even though we do say *very profitable property*; thus, the ability to co-occur with degree adverbials has **not** been generalized from predicating to nominal adjectives.) A "generalization" in this sense simply describes one kind of troublesome fact that must still be explained in some other way. Thus, despite the intuitive plausibility of treating the sentences in (7.5) as a rule generalization, this cannot be regarded as an argument or explanation of any real substance.

A far more satisfying explanation of the data of (7.5) is available, however, if we hypothesize that these adjectives are derived by **ellipsis** from their usual prenominal position within normal CNs. That is, rather than arising like true predicating adjectives from predicate position in semantic structure, these adjectives are formed by regular CN formation processes and are subsequently left stranded when their head nouns are deleted under identity by a very late rule. The steps that such a derivation would take are indicated informally in (7.6).

(7.6)   a. *That interpretation is an interpretation by a president.*
That interpretation is a president's interpretation.
That interpretation is a presidential interpretation.
That interpretation is presidential.
   b. *His razor is a razor using electricity.*
His razor is an electricity-using razor.
His razor is an electricity razor.
His razor is an electric razor.
His razor is electric.
   c. *Her infection is an infection caused by a virus.*
Her infection is a virus-caused infection.
Her infection is a virus infection.
Her infection is a viral infection.
Her infection is viral.

The essential point here is that the derivations suggested in (7.6) derive the adjectives which appear in predicate position on the surface from an earlier stage in which they are **prenominal** modifiers within a CN. Their appearance in predicate position is therefore a result of the fact that each CN within which they are derived is itself the head of an NP that happens to occupy predicate position. This contrasts directly with the usual derivation for predicating adjectives, which involves no such prior stage; on the contrary, the appearance of any predicating adjective in prenominal position follows, rather than precedes, its appearance in predicate position.

One important advantage of this approach is that it requires that the derivation for either of the two sets of adjectives be consistent, regardless of where in the larger sentence the adjective originates. Specifically, nominal adjectives would arise only within full CNs whether or not that CN was itself part of a predicate nominal, whereas predicating adjectives would be derived as independent constituents, whether or not they were eventually to be preposed. As a consequence, the syntactic derivation of an adjective will be predictable solely on the basis of its semantic structure (i.e., whether it is semantically equivalent to a noun or not) and not at all in terms of its eventual position in surface structure.

My claim that the adjectives that surface in predicate position are of two fundamentally different sorts, and hence must have followed systematically distinct derivational paths to arrive at the same point, is supported by at least three different kinds of syntactic evidence relating to the distribution of these adjectives in predicate position. The first of these concerns the range of grammatical subjects that we find in sentences where each of the two adjective types appears in predicate position. Some of the relevant data, based on the CNs *theatrical agent* and *chemical engineer*, are shown in (7.7). (Note that the nominal adjective *theatrical* —derived directly from *theater*—must not be confused in any of these examples with the homophonous predicating adjective that means, roughly, 'extravagantly dramatic'.)

(7.7)   a. $\left\{\begin{array}{l} \textit{Our engineers} \\ \textit{*Those agents} \\ \textit{*My relatives} \end{array}\right\}$ *are all chemical.*

        b. $\left\{\begin{array}{l} \textit{Those agents} \\ \textit{*Our engineers} \\ \textit{*My relatives} \end{array}\right\}$ *are all theatrical.*

        c. *That woman, who is an engineer, is* $\left\{\begin{array}{l} \textit{bright} \\ \textit{*chemical} \end{array}\right\}$.

        d. *That guy, who is an agent, is* $\left\{\begin{array}{l} \textit{talented} \\ \textit{*theatrical} \end{array}\right\}$.

In a theory that treats all predicate adjectives as being derived in the same way, we would have no straightforward explanation for the fact that the adjectives shown in (7.7) can grammatically co-occur with quite different sets of subjects. In contrast, the theory proposed here can explain the distribution of asterisks in these examples on the basis of the two derivational paths outlined earlier. (This entails, of course, that the only adjectives that can originate as constituents of the highest S are predicating ones.)

In particular, the derivation by ellipsis proposed here for nominal adjectives that appear in predicate position can account directly for these grammaticality judgments. Since the head noun that I claim is elided from predicate position can only be so elided under conditions of identity with a preceding occurrence of the same noun as head of a clausemate subject (e.g., *Those agents are all theatrical* from *Those agents are all theatrical agents*), it follows that sentences that lack such an antecedent should be ungrammatical with nominal adjectives appearing alone in predicate position; a check of the data in (7.7) reveals that the asterisks assigned there correspond precisely to such cases.[4] (Note that the presence of appropriate antecedents in nonrestrictive relative clauses, such as *engineer* in [7.7c] and *agent* in [7.7d], is insufficient to achieve grammaticality since the clausemate condition in the structural description for ellipsis is not met.)

There is a second way (not unrelated to the first) in which the derivations proposed here predict the grammaticality facts of (7.7). The claim that nominal adjectives arise only within CNs accurately predicts that the range of subjects that can co-occur with nominal adjectives as superficial predicates will be just those NPs whose head nouns form acceptable CNs with the same adjectives.[5] Thus, the distribution of asterisks in (7.7a) and (7.7c), for example, is predictable from the fact that *chemical engineer* is a "good" CN, whereas *chemical agents* (taking *agent* in its animate sense), *chemical relatives*, and *chemical woman* are not.[6]

---

[4] Georgia Green has brought to my attention (personal communication) that this hypothesis also provides an explanation for the fact that both the plausibility of deriving nominal adjectives in predicate position by ellipsis from full CNs and the systematic synonymity manifested by the corresponding pair of sentences do not carry over at all to comparable cases involving predicating adjectives. Thus, although (i) is at least a plausible source for (ii), there is **no** single comparably plausible source with a full NP in predicate position that could be systematically associated with sentences such as (iii), (iv), and (v).

  (i) *That hypothesis is an absurd hypothesis.*
  (ii) *That hypothesis is absurd.*
  (iii) *Ishmael is obese.*
  (iv) *They are all exceptionally cantankerous.*
  (v) *No one is immune.*

This discrepancy between sentences containing nominal and predicating adjectives, however, follows naturally from the derivation-by-ellipsis hypothesis proposed here, and thus provides additional support in its favor.

[5] Note that it would be inaccurate to speak of subject Ns or NPs which are "modified" by these nominal adjectives, since the latter only **appear** to be modifying the grammatical subjects as a result of the ellipsis. Semantically, the only noun which the surviving nominal adjective may be said to modify is the noun with which it formed a CN prior to the ellipsis, while it is that entire CN which itself modifies the subject (to the extent that predicate nominals may be said to "modify").

[6] My claim that *chemical agents* is not a "good" CN is valid only under the assumption

The second piece of syntactic evidence that supports a systematic distinction among the adjectives in question consists of an additional restriction on the use of adjectives in predicate position; here too, the restriction applies to nominal adjectives but not to predicating ones. The specific constraint is that nominal adjectives may not appear with semantically indefinite subjects.

As the reader may easily verify, all the sentences with nominal adjectives in predicate position cited thus far in (7.5)–(7.7) contain **definite** subject NPs. That this is not an accidental distribution may be seen from the data in (7.8), whose relative clauses were deliberately constructed to show indefinite subjects. In each case, the nominal adjective is grammatical in prenominal position (i.e., within a CN) but ungrammatical when used alone in predicate position in a corresponding relative clause with an indefinite subject.

(7.8)   a. *I wish I had some* $\begin{cases} \textit{musical talent} \\ \textit{*talent that was musical} \end{cases}$.

        b. *Rita wants to edit a* $\begin{cases} \textit{linguistic journal} \\ \textit{*journal which is linguistic} \end{cases}$.

        c. *We're tempted to press* $\begin{cases} \textit{criminal charges} \\ \textit{*charges that are criminal} \end{cases}$.

        d. *The prize money is for* $\begin{cases} \textit{regional novelists} \\ \textit{*novelists who are regional} \end{cases}$.

These data present a clear and consistent contrast with parallel examples using predicating adjectives, as may be seen in (7.9):

(7.9)   a. *I wish I had some* $\begin{cases} \textit{marketable talent} \\ \textit{talent that was marketable} \end{cases}$.

        b. *Rita wants to edit a* $\begin{cases} \textit{fashionable journal} \\ \textit{journal which is fashionable} \end{cases}$.

        c. *We're tempted to press* $\begin{cases} \textit{very serious charges} \\ \textit{charges that are very serious} \end{cases}$.

        d. *The prize money is for* $\begin{cases} \textit{unknown novelists} \\ \textit{novelists who are unknown} \end{cases}$.

---

that *agent* has the same meaning as in forms like *theatrical agent, literary agent*. If, however, it is used in a sense something like 'salesperson' so that *chemical agent* would be parallel to other forms such as *hardware agent, glassware agent*, or perhaps *insurance agent*, it would be a viable CN. This, however, does not invalidate the point in question here, since it remains true that the grammaticality and semantic well-formedness of a sentence such as *Those agents are chemical* are predictable precisely from the grammaticality and well-formedness of *Those agents are chemical agents*, as I have claimed.

Once again, a theory in which nominal adjectives in predicate position are derived by ellipsis can provide an explanation for what would otherwise be a very puzzling contrast. In such a theory, CNs like *musical talent* or *linguistic journal* are derived from intermediate constructions comparable to *talent for music* and *journal for linguistics*, respectively. Appropriate deriviations for the acceptable sentences of (7.8a) and (7.8b) would therefore follow the steps sketched in (7.10) and (7.11), respectively.

(7.10)　　*I wish I had some talent that was for music.*
　　　　　I wish I had some for-music talent.
　　　　　I wish I had some music talent.
　　　　　I wish I had some musical talent.

(7.11)　　*Rita wants to edit a journal which is for linguistics.*
　　　　　Rita wants to edit a for-linguistics journal.
　　　　　Rita wants to edit a linguistics journal.
　　　　　Rita wants to edit a linguistic journal.

However, since I claim that nominal adjectives may emerge in predicate position only when stranded there by ellipsis, my theory would require derivations like those in (7.12) and (7.13) to generate starred sentences like those of (7.8); note in particular the boldfaced words, for they will be seen to be the source of the ungrammaticality of the final output.

(7.12)　　*I wish I had some talent **that was** [**talent** that was for music].*
　　　　　I wish I had some talent that was [for-music talent].
　　　　　I wish I had some talent that was [music talent].
　　　　　I wish I had some talent that was [musical talent].
　　　　　**I wish I had some talent that was musical.*

(7.13)　　*Rita wants to edit a journal **which is a** [**journal** which is for linguistics].*
　　　　　Rita wants to edit a journal which is a [for-linguistics journal].
　　　　　Rita wants to edit a journal which is a [linguistics journal].
　　　　　Rita wants to edit a journal which is a [linguistic journal].
　　　　　**Rita wants to edit a journal which is linguistic.*

The reason that these outputs (like the others marked with asterisks in [7.8]) are ungrammatical is that the underlying structure is semantically ill-formed, by virtue of the useless redundancy conveyed by the portions shown in boldface. Thus, by providing a well-motivated source for nominal adjectives that appear grammatically in predicate position, the derivation proposed here can apparently also predict the lack of grammaticality in partly parallel sentences like those marked with stars in (7.8).

The third syntactic fact that distinguishes nominal adjectives from predicating adjectives in predicate position is that the former are consistently and markedly more acceptable when used in an explicit or implied comparison than when they are used alone. Thus, the (a) sentences in (7.14)–(7.17) are regularly more acceptable to me than the (b) sentences, which lack a contrasting adjective.

(7.14) *Her infection turned out to be*
    a. *viral, not bacterial.*
    b. *viral.*

(7.15) *The strongest drives toward pollution control have been*
    a. *governmental rather than industrial.*
    b. *governmental.*

(7.16) *Our firm's engineers are all*
    a. *mechanical, not chemical.*
    b. *mechanical.*

(7.17) *That interpretation of the subpoena is*
    a. *presidential, not judicial.*
    b. *presidential.*

Not surprisingly, these adjectives also become more acceptable when preceded by qualifying adverbs such as *primarily, mainly, mostly,* or *largely* since these modifiers function in such a way as to set up an implied comparison, as may be seen in (7.18).

(7.18) a. *The therapy he does is* $\left\{ \begin{array}{l} \textit{primarily musical} \\ \textit{?musical} \end{array} \right\}$.

    b. *The novelists we studied were* $\left\{ \begin{array}{l} \textit{mostly regional} \\ \textit{?regional} \end{array} \right\}$.

    c. *The equipment they sell is* $\left\{ \begin{array}{l} \textit{mainly culinary} \\ \textit{?culinary} \end{array} \right\}$.

In contrast, when we use regular predicating adjectives alone in predicate position, we experience no sense at all that we **must** add a contrasting adjective in order to improve well-formedness; thus, the sentences that follow in (7.19) are perfectly acceptable without any explicit or implicit contrast. Moreover, even though a contrasting expression may be added without destroying grammaticality (as the phrases in parentheses show), such additions often seem to add more anomaly than clarification.[7]

---

[7] One likely source of the anomaly of these parenthetical additions is the fact that they add no new information to what the first adjective already conveys (at least to a native speaker of the language). Interestingly enough, this provides an explanation of the related

(7.19) a. *Her infection turned out to be serious (not light).*
  b. *The government's moves toward pollution control were popular (not despised).*
  c. *The engineers in our firm are all young (not decrepit).*
  d. *That interpretation is untenable (not justifiable).*

This third syntactic distinction is, on the face of it, more puzzling than the first two since it is difficult to see why the fundamental difference in predicability that distinguishes the two sets of adjectives should set up a concomitant difference in the suitability of a syntactic structure with explicitly contrasted adjectives; moreover, the hypothesis advanced earlier which sets up two distinct derivational tracks for the two adjective types does not seem to make direct predictions of any sort about acceptability in a contrastive frame. Nevertheless, the syntactic difference shown by (7.14)–(7.19) is so clear and consistent that it even permits us to "rehabilitate" some of the starred sentences seen earlier in (7.8) by using negation to set up **implied** contrasts, like those shown here:[8]

(7.20) a. *I wish I had some talent that wasn't musical.*
  b. *Rita wants to edit a journal which isn't linguistic.*
  c. *?The prize money is for novelists who are not regional.*

One possible explanation is suggested by some remarks made by Bartning (1976:75–80) in connection with French nominal adjectives appearing in predicate position. She explains the suitability of a contrastive frame for nominal adjectives in terms of the different kinds of semantic oppositions set up by nominal adjectives as compared to predicating ones, namely, that predicating adjectives typically name a point along a continuum (*l'opposition polaire*), whereas the nominal adjectives used in CNs typically denote one of a number of "appropriate" subsets (*l'opposition multiple*, whose limiting case is simply *l'opposition binaire*).[9] These distinctions are illustrated in (7.21)–(7.22), which show contrastive structures for the predicating and nonpredicating readings of the ambiguous expressions *civil engineer* and *diplomatic historian*; in each case, the contrastive frame is rather jarring for the predicating sense of the adjective shown in (a) but perfectly acceptable for the nonpredicating sense shown in (b).

---

fact that such additions could become extremely functional, and thus lose their anomaly, in a somewhat different conversational context, namely, if they were used as definitional paraphrases (in lieu of direct translation) in a conversation with a non-native speaker unlikely to recognize the full meaning of the first adjective. (One tends to forget that such conversations must also represent perfectly well-formed stretches of English discourse.)

[8] This use of negation was brought to my attention by J. McCawley.

[9] See Section 2.2.1 for further discussion of these three fundamental oppositions, which Bartning bases on Leech 1974:106ff.

(7.21)   a. *?That engineer is civil, not rude.*
         b. *That engineer is civil, not biomedical.*

(7.22)   a. *?That historian is diplomatic, not abrasive.*
         b. *That historian is diplomatic, not social.*

It is apparent from these data that the addition of a contrasting adjective strikes us as significantly more redundant when it follows a predicating adjective than when it follows a nonpredicating one. Bartning suggests, quite reasonably, that the redundancy in the first case stems from the fact that the specification of one point along a continuum (e.g., the "civil" point along a continuum of politeness) necessarily excludes all other points, including the one named by the contrastive adjective. In the second case, however, the fact that nominal adjectives typically name one of a number of subsets that stand in a semantic relation of "multiple opposition" to one another (e.g., *civil engineers* as opposed to other more or less discrete subsets such as *mechanical engineers, biomedical engineers, aeronautical engineers,* and so forth) renders the contrastive syntax shown in the preceding examples, much more natural, and perhaps even anticipated (Bartning 1976:80).

It thus appears that the syntactic–semantic distinction represented by the examples of (7.14)–(7.17) on the one hand, and those of (7.19) on the other, has a natural semantic explanation in terms of the types of oppositions typically established by the two kinds of adjectives in question.

I have argued in this section that the appearance of nominal adjectives in predicate position, although somewhat unexpected, may best be viewed as the consequence of a late rule of ellipsis which leaves the prenominal adjective of a CN stranded after the copula. The derivation of these adjectives is therefore no different from their derivation in cases where ellipsis does not apply and, in particular, is systematically distinct from the derivation of predicating adjectives that surface in the same syntactic slot.

7.3 Analytic Indeterminacy and Individual Variation

The major emphasis in this work has been on identifying and describing the most regular and systematic aspects of the grammar of complex nominals; to this end, specific derivations have been proposed and a wealth of syntactic evidence amassed to support each of the hypotheses which together constitute the present theory. However, the fact that systematic regularities are discernible and that detailed derivations may be worked out in a coherent fashion should not blind us to another side of this area of the grammar where details are fuzzier and the significant variables harder to untangle. Two aspects of this "fuzzier" side of the grammar of CNs to be taken up in this section are a certain "indeterminacy of analysis" which resists reduction to absolute statements, and individual variation and its effects on our understanding (and our description) of CNs in natural language.

The "indeterminacy of analysis" at issue here refers to the fact that for a certain portion of the CN forms, there seems to be no clear and principled way to decide under which RDP (or nominalization) heading they should be listed; that is, given a specific surface form and its familiar referent, more than one LS suggests itself as an appropriate underlying structure. (Restated in terms of the language user's point of view, one finds disagreement among different speakers concerning "the" meaning of a particular surface CN, even when the same [familiar] referent is intended.) For instance, there is a set of CN forms in English that name both artifacts and natural masses in terms of their constituent material; examples include *copper coins*, *bronze statue*, and *sugar cube* on the one hand, and *sand dune*, *raindrop*, and *land mass* on the other. In this work, these forms are listed under both MAKE₂ and BE, on the grounds that each of these analyses seems well motivated intuitively and grammatically (i.e., we find surface reflexes of both underlying MAKE and underlying BE)[10] and there seems to be no principled reason to prefer one over the other.

The major question that such indeterminacy poses for my theory is whether it is compatible with a strict claim of multiple ambiguity such as the one that has been advanced throughout this book.[11] Ambiguity, after all, is generally viewed as the consequence of having a number of underlying structures converge into the same surface representation. The problem is, however, that if we cannot identify **for certain** just what the underlying structure must be for each particular reading of a surface CN, then the posited ambiguity which is an essential component of this analysis may dissolve into a hazy cloud of derivational uncertainty. Or so it first appears.

I believe, however, that the demand which is implicit in the preceding criticism to the effect that a theory of CNs must associate only **one** LS with a given CN (used to denote a certain referent) is fallacious on a

---

[10] Thus, we may say either *Those coins are copper* (BE) or *Those coins are made of copper* (MAKE); or we may elicit the answer *Land* to either of these two questions: *What is that mass?* (BE), *What is that mass made (up) of?* (MAKE).

[11] I am grateful to Jerry Sadock for bringing this apparent conflict to my attention. Parts of the discussion to follow, in which I attempt to answer his objections, have appeared previously in Levi 1976, Section 5.1.

number of grounds. To begin with, it is likely that some of the more pervasive "double analyses" suggested by the data may be a function of interrelationships among different RDPs which are predictable on independent semantic grounds. For example, it is by no means arbitrary that expressions like *copper coins* or *chocolate bunny* may be analyzed under either BE or MAKE₂, since the fact that an item is made out of a certain material is often expressed in natural languages by saying that the item "is" that material; this is precisely why we can say both *That bunny is chocolate* and *That bunny is made of chocolate* in speaking of the referent of the CN *chocolate bunny*.

Similarly, it is not accidental that a CN such as *party members* may be analyzed under either HAVE₂ or IN since each relation entails the other; it is a function of the meaning of the modifying noun *party* (in the political sense) that the members that it **has** are said to be contained **within** it, or conversely, if there are members **in** the party, then the party **has** members. (Similar entailments may be observed for CNs whose nouns all have concrete referents, such as *ocean floor*, *body fluids*, *brain cells*, *city parks*, or *library stacks*.) In these cases, and quite probably a number of others, there is simply no principled basis (that I am aware of) for assigning cognitive primacy to one of the two intertwined relationships and, hence, to one derivational type over the other. To be sure, these semantic entailments must be spelled out explicitly and supported by the evidence of grammatical reflexes, but such an assignment is by no means an impossible one to carry out; moreover, attempts to do so are quite likely to enrich our understanding of the universal principles involved in CN formation, since the nature of these "double analyses" suggests clearly that they are not a language-specific phenomenon. It is essential to observe, however, that in the cases just cited and all others of which I am aware, the apparent indeterminacy represents a choice among readings predicted by the present theory, so that the **range** of semantic variation we would normally expect for CN forms is in no way violated. We may conclude, then, that for this kind of indeterminacy, the fact that no single LS may be assigned to a given CN denoting a given referent is explainable on independent semantic grounds.

A different reason for rejecting the demand that a theory of CNs must assign a unique semantic structure for every CN form paired with a specific referent is that it assumes that all speakers, in associating a given N–N form with a particular real world object, will make the same cognitive connections and thus judge the same relationship to be the most salient. I suggest that this assumption rests on just those "two very naive, epistemological fallacies" which Reddy (1969) has described in the following manner in the context of understanding metaphor:

First, to embody in a grammar formally those restrictions which seem to operate in the normal, external physical world is to assume that all human utterances are directly and primarily concerned with describing this world. Second, to embody in a grammar those restrictions which seem to operate in the normal, external physical world is to assume that there is some normal, external physical world from which to extract them [p. 243].

In the context of CNs, this quotation suggests that it is erroneous to assume that the primary purpose of associating a surface CN with a particular referent is to describe it as directly or as accurately as possible; if this were our major intent, surely a less abbreviated, more explicit paraphrase would enhance our descriptive precision. As suggested earlier, however, it is far more likely that the basic motivation in using a particular CN is the highly pragmatic one of selecting the most concise linguistic form that will still ensure that both speaker and listener end up talking about the same object; for this purpose, achieving total agreement on the underlying structure of the CN may be more serendipitous than indispensable.

Reddy's remarks also suggest that it is foolish to expect the real world to provide a single answer to the question, "Why is this object named by these two nouns?" since their adjunction in a CN is not a direct function of the physical world but rather of the ways in which speakers perceive it—ways that are certain not to be uniform. Thus, if one asks English speakers why a tidal wave is called a *tidal wave*, one gets different answers; some people say, "Because it is caused by the tide" (a belief which, though counterfactual, seems widely held), while others answer, "Because it sweeps in like the tide, only it's more powerful." Similarly, if one asks what *job tension* means, some speakers will offer a causative paraphrase such as "tension caused by your job" whereas others will assign a simple locative reading instead, such as "tension you feel on the job."

While such intersubjective differences are quite common for high frequency CNs which, in a sense, we all inherit without explicit explanation (be it etymological or synchronic), these differences are also likely to crop up when novel CNs are introduced. For example, I recall being puzzled once by a sign in the campus bookstore reading *Athletic Charges*, a CN I had never before encountered. Although the clerk provided a very full explanation to the effect that students on athletic scholarships had their book bills charged to the Athletic Department, her description did not provide any direct way for me (*qua* curious linguist) to decide whether the CN should be derived by FOR Deletion from *charges for athletes* or by a subjective nominalization process from *acts of athletes' charging [something]*. Knowledge of what a CN refers to simply does not entail certain knowledge of its derivation, if by "its derivation" we must assume a unique path. (For that matter, the assumption that all speakers could ever reach agreement on "the" definition of **any** term, regardless of its syntactic or semantic complexity, is certain to be fallacious as well, particularly in view of the two very pertinent observations made by Reddy in the quotation cited earlier.)[12]

What **is** essential to verify, however, is whether the different readings associated with a given CN by different speakers, as a function of their own perceptions of and beliefs about the world, fall within the range of readings that is predicted by the present theory. Given the semantic flexibility which is inherent in the multiple ambiguity of CN forms, this seems to be the strongest demand that we can make of the theory in terms of predictive power.

The problem raised here of the ways in which speakers differ in their use and interpretation of CNs is a complex one which has only begun to be explored by contemporary linguists. Some of the questions that invite more extensive study include (*a*) the range and nature of individual variation both in "tolerance" for unfamiliar CNs (an issue raised in Zimmer 1972:17 and in Gleitman and Gleitman 1970:115–118, 137–141), and in creativity in introducing new ones; and (*b*) the influence of cultural and social variables on the degree of transparency presented by specific sets of CNs (e.g., which groups of speakers understand the two ways in which engineering subfields are named, and which groups use terms like *biomedical engineering* or *chemical engineering* as unanalyzed wholes?). In this context, one might also investigate Zimmer's claim (1972) that novel CNs are bestowed as names only on those entities that particular speakers judge to be "appropriately classificatory" in their own world. Since it seems reasonable to expect that such judgments would be correlated with social variables such as age, education, and occupation, an investigation of such variables might help to make more precise just what factors this notion of "classificatory relevance" is really based on.

Although studies of this sort necessarily entail many problems of both a theoretical and a methodological nature, the attempts to resolve such problems, as well as the results eventually obtained, may well prove illuminating for an overall theory of complex nominals.

The last area of fuzziness to be examined here is one in which the problem of individual variation impinges rather directly on our theoretical description; the specific question is where and how a grammar is to draw a line between lexicalized and nonlexicalized CNs. I have claimed that my theory can account for the semantic structures and derivations of only those CNs that may be productively derived by native speakers, and cannot be expected to account for the semantic structures of those forms

---

[12] See also Reddy 1973 and 1977 for instructive elaborations on this theme.

that have become lexicalized. I also claimed (Section 1.3) that this distinction is not a circular one but rather one which makes empirically testable claims about ease of acquisition by both children and non-native speakers, directness of translation into other languages, and suitability for extension of the underlying pattern to produce novel forms.

The first problem that arises in considering such a distinction, however, is the fact that one person's transparent (derived) form is another one's lexical item; the second, and related, problem is how the grammar is to account for such forms. Must it generate all forms that are transparent to at least some speakers? Must it list in the lexicon all forms that are opaque (i.e., nonderived) for at least some speakers? Should it do both simultaneously, and if so, should it relate the two accounts in some way? (Or, for that matter, are all these questions dependent on the more fundamental one of whether it makes sense to speak of "the" grammar at all?)

The challenge that such questions pose to a theoretical description of CNs is closely comparable to the dilemma addressed by Zimmer (1964) in his study of affixal negation as "an investigation of restricted productivity." Faced in that context with the similar issue of which adjectives in *un-* should be grammatically derived and which should be lexically listed, Zimmer suggests a compromise solution according to which the lexicon would list a number of these adjectives "on a par with monomorphemic forms" while the grammar would also include rules for generating those forms that represent the most **productive** (i.e., "unrestricted") patterns. He then explains his position in this way:

> The implication of such an approach as a model for the linguistic behavior of speakers of English is that a number of forms such as *untrue, unhappy, unkind* are learned as lexical items like *true, happy, kind* while other forms in *un-* can reasonably be accounted for as the output of productive rules. . . . This does not mean that individual forms belonging to this second group may not often be acquired by repetition; what we are concerned with is the readiness with which speakers do use forms that they may never have heard before, and the readiness with which such forms are accepted. . . .
>
> What we claim is, in effect, that a purely generative account of all morphologically complex forms may lead in certain cases to an unrealistically simple description, and that we need a grammar which will *analyze* such morpheme combinations but will not *generate* them [Emphasis added; pp. 85–86].

These remarks of Zimmer's raise a number of distinct but related issues whose relevance extends beyond the affixation data with which he was primarily concerned, to the problem in question here, namely, the distinction between lexicalized and nonlexicalized CNs and their appropriate treatment in a grammatical description. The issues relevant to both contexts include these: (*a*) recognizing that morphologically similar forms may need to be accounted for in two quite distinct ways (i.e., grammatical derivation versus lexical listing); (*b*) distinguishing between the ways in which we learn certain forms and the ways in which a grammar should represent them; and (*c*) the possibility of including in a grammar both a "generative" and an "analytic" morphology (or some other equivalent mechanism) to correspond to the distinction between truly productive word formation processes and those patterns with radically restricted (or even zero) productivity.[13]

All of these issues are challenging ones, not least because they suggest that an adequate account of CN formation (along with other word formation processes) in natural language may entail more areas of fuzziness than have hitherto been recognized by transformational grammarians in particular. For example, the question of just when we may say that a particular CN form has "crossed over" into the lexicon may simply have no clear-cut answer, in view of the fact that its "definition" and thus its derivational transparency are certain to vary from speaker to speaker.

Another area in which our investigations may turn up no satisfyingly sharp and neatly formalizable results is that of language acquisition; specifically, we need to ask such questions as how and when children learn the productive rules of CN formation,[14] whether they go through stages in which their rules are not only simpler than the adult's but significantly different (e.g., whether they ever use **more** RDPs than adult speakers would), and whether older children "go back" and reanalyze CNs they once learned as units so as to achieve a simpler, less redundant mental grammar of CNs.[15]

Finally, we must ask ourselves what implications for the overall theory are carried by the dual morphological system (i.e., "generative" plus "analytic") suggested by Zimmer. It may be, for example, that our traditional expectation that a rigid separation can be maintained between the outputs of "grammatical processes" like those described earlier in Chapters 4 and 5, and the idiosyncratic entries assigned to the lexicon, is just not a realistic one, **especially** in the area of word formation where syntactic regularities and lexical idiosyncrasies seem to mingle in a rather

---

[13] See Thompson 1974 for a useful discussion of these and other important issues relating to the general problem of lexical productivity.

[14] The only two studies that I am aware of that deal explicitly with this question are Livant 1962 and Oestman 1972, although both are unfortunately very limited in scope.

[15] McCawley (1976d) suggests not only that this sort of "optimizing" is unlikely to take place in a child's grammar (especially in the area of morphological acquisition), but also that the nonoptimal and idiosyncratic ways in which we acquire morphologically complex forms (including CNs) may be the major reason for the well-known individual variations found in adult speakers with regard to morphological analysis of specific complex forms.

  7 EXCEPTION CLASSES AND OTHER PROBLEMS

unorthodox fashion. In such a case, a two-track system with a certain amount of overlap may actually be more descriptively adequate than a theory which carries at least the implicit claim that every CN form used or understood by a given speaker must be classified either as "generated" or as "learned" but not as both.

The major thrust of this work has been to demonstrate the pervasive regularities that may be discerned in the area of CN formation. The major purpose of this section, on the other hand, has been to call the reader's attention to those aspects of CN formation that resist characterization in neat and easily formalized terms. In a sense, then, these remarks may be construed as an invitation to the reader to share with me both my sense of chagrin and my sense of excitement at having discovered, for the nth time, how easy it is to underestimate the full complexity of any natural language phenomenon.

8

SUMMARY AND CONCLUSIONS

all ignorance toboggans into know
and trudges up to ignorance again
— e. e. cummings[1]

8.1 Summary

The primary purpose of this book has been to explore the syntactic and semantic properties of the construction called the complex nominal, and to propose appropriate underlying structures and transformational derivations within generative semantic theory. In addition, an attempt has been made to situate the description within the broader framework of universal grammar, and to identify those areas for which this study has relevance outside the domain of linguistic theory per se.

In Chapter 1, relevant examples of complex nominals were introduced, drawn from data previously studied under the separate headings of (*a*) compound nouns, (*b*) nominalizations, and (*c*) nonpredicating adjectives within nonpredicate NPs. The claim was advanced that all these forms represent a single type of grammatical constituent, called the complex nominal or CN, whose underlying structure contains both a nominal head and a complement S and whose surface structure must be dominated by a node label of N. In addition, it was hypothesized that all CNs are derived by just these two syntactic processes: predicate deletion

---

[1] From e. e. cummings, *Complete Poems 1913–1962*. New York: Harcourt Brace Jovanovich, Inc., 1962. Reprinted by permission of the publishers.

and predicate nominalization. In connection with the first of these, the notion of the Recoverably Deletable Predicate (RDP) was introduced along with the claim that a limited and specifiable set of such RDPs can account for the multiple semantic structures that may regularly be associated with CNs derived by predicate deletion. A number of restrictions on the scope of the inquiry were then made explicit, and a small group of new terms and abbreviations introduced.

In Chapter 2 the hypothesis of nominal origins for nonpredicating adjectives within complex nominals was explained, and arguments based on a variety of syntactic and semantic phenomena were provided in its support; the scope of the work was then restated in the light of the preceding sections. In the concluding section of the chapter, it was demonstrated that the set of complex nominals encompasses a broader range of data than that associated with the "nominal compound" or "compound noun" of traditional studies; more significantly, it was also shown that no clear and consistent definition of this latter entity has ever been provided in either traditional or transformational studies. As a result, the existence of a grammatically significant subset of CNs that would correspond to the traditional term of "nominal compound" or "compound noun" was explicitly denied.

A broad foundation for the theory was established in Chapter 3, where a number of general issues pertaining to complex nominals were explored in some detail. The central issue of whether CNs are to be treated as primarily idiosyncratic (i.e., lexical) units or as multiply ambiguous configurations amenable to formal analysis was discussed, and reasons were provided for rejecting the former approach as inherently inadequate. The distinction between "familiar" and "attested" CNs on the one hand, and "grammatically possible" CNs on the other, was then drawn in order to delineate the objectives of the study more precisely. An examination of the communicative function served by complex nominals followed, in which the characteristic conciseness of CNs and their suitability to serve as "ad hoc names" were emphasized. Finally, systematic evidence was presented in support of the contention that all CNs must be analyzed syntactically as Ns rather than as NPs.

Chapters 4 and 5 then provided the specific derivational details that constitute the core of the theoretical analysis. The derivation of CNs by predicate deletion was the subject of Chapter 4, in which the idea of a set of Recoverably Deletable Predicates (RDPs) plays a central role. This idea amounts to the claim that all CNs not derived by nominalization processes have their origins in semantic structures from which only these nine predicates may be subsequently deleted: CAUSE, HAVE, MAKE, USE, BE, IN, FOR, FROM, and ABOUT. Following some observations about the overall composition of this set, each of these predicates then received explicit attention and justification. Full derivations of three major CN types were proposed, and the crucial transformational processes of Compound Adjective Formation, Recoverable Predicate Deletion, and Morphological Adjectivalization were analyzed in detail. After a summary of the evidence that may be adduced in support of these derivations and processes, and a discussion of listener strategies for disambiguating surface CNs, the chapter concluded with some external evidence from English and other languages suggesting that this area of the grammar reflects a variety of universal principles and constraints.

Chapter 5 dealt with the derivation of CNs by predicate nominalization. A new analysis of English nominalizations was proposed in which nominalized verbs were divided into four fundamental types: Act, Product, Agent, and Patient Nominalizations. CNs in which such nominalizations form head nouns were then classified into two major groups, namely, Subjective and Objective NOM CNs. The crucial rules by which the nominalization heads are formed, as well as full derivations for all NOM CN types, were then presented and analyzed in detail. The chapter concluded with a section providing arguments in support of the cyclicity of the crucial rules in these derivations, based largely on the recursive nature of the process and on the interaction of the two sets of rules in Chapters 4 and 5.

Ramifications of the theory beyond its original domain were taken up in Chapter 6, where questions were discussed pertaining to stylistic variation in surface CN forms, appropriate lexicographic treatment of CNs and nominal adjectives, the interaction of semantic and pragmatic principles in our use of CNs, and the relation of this study to universal grammar. Chapter 7 then served to introduce a number of potential and actual problems for the theory, including several sets of data unaccounted for by the rules of Chapters 4 and 5. Explanations for all but one of these data sets were suggested, along with a hypothesis based on ellipsis to explain the occasional (and somewhat unexpected) appearance of nominal adjectives in predicate position. In the last part of Chapter 7, analytic indeterminacy and intersubject variation were discussed as two aspects of the grammar of complex nominals which offer challenges to the present theory.

8.2 New Answers, New Questions

In the first chapter of this book, it was pointed out that although the topic of so-called nominal compounds has received extensive attention over the years in studies of English and other languages, the subject of nominalizations has been less thoroughly investigated; moreover, the very existence of nonpredicating adjectives, as well as their significance for linguistic theory, had been almost totally ignored until the appearance of Bolinger's perceptive comments on these forms in a 1967 article in *Lingua*. The integration of these three formerly disparate topics into a single coherent theory of complex nominals must therefore be regarded as one of the most important contributions made by this study to our understanding of this area of the grammar.

An analysis that provides a unified treatment of previously unrelated phenomena is worth little, however, unless the attempts at integration are motivated at every step by systematic argumentation and extensive empirical evidence. In accordance with this requirement, therefore, I have tried throughout this book to provide supporting evidence not only for the novel hypotheses introduced here for the first time, but also for a number of claims made in earlier published analyses which, while pertinent, had more often been treated as self-evident assumptions than as hypotheses requiring explicit and systematic justification. As a result, this work provides a more comprehensive and fully argued analysis of complex nominals than any other of which I am aware.

A number of theoretical innovations have been introduced in these pages. Among the most important of these must be counted:

1. The explanation provided for the unusual behavior of nonpredicating adjectives by analyzing them as essentially stylistic variants of modifying nouns within CNs;
2. The claim that all CNs must be derived by either predicate deletion, predicate nominalization, or a combination of both;
3. The proposal that a small and specifiable set of Recoverably Deletable Predicates can account for the multiple ambiguity of productively formed CNs that are not produced by nominalization;
4. The analysis of nominalizations in English according to which four basic types of nominalized verbs are derived from two basic types of underlying structures;
5. The integration of the two distinct derivational types of CNs (namely, the RDP Deletion type and the NOM CN type) into a single grammatical description which makes explicit both their similarities and their differences.

Although I have been primarily concerned with providing a well-motivated transformational analysis of the syntactic and semantic regularities of complex nominals, I have also tried to relate my description of these aspects of CNs to other contemporary studies (namely, those of

Zimmer and Downing) which emphasize the **functional** and **pragmatic** aspects of complex nominals in human communication. Although the links established among these areas (especially in Chapters 3 and 6) are somewhat preliminary in nature, it is my hope that these suggestions will provide constructive guidelines for more elaborate consideration of these issues in the future.

Another underexplored area for which this book has relevance is that of word formation processes in natural language, an area of the grammar which is only now beginning to re-emerge into the analytic sights of linguists from its rather prolonged period of neglect by transformational grammarians in particular. Here too I hope that this study of complex nominals will provide insights of more general value in illuminating such problematic—and challenging—issues as the alleged boundary between morphology and syntax, and the distinctions to be made between lexical and sentential regularities.

Still another way in which the present analysis extends beyond the confines of its original domain is in its implications for universal grammar. To begin with, we have seen that the nine RDPs originally motivated to account for English CNs alone appear to constitute a set that has universal linguistic significance on both syntactic and semantic planes. Moreover, an examination of some of the semantic and pragmatic principles which guide English speakers in using English CNs suggests that these principles are also likely to reflect linguistic universals that constrain the use and interpretation of such forms in all human languages. As a result, the analysis proposed here receives additional support on the grounds of external adequacy, while at the same time it suggests a variety of rich and promising themes for continuing crosslinguistic research.

This book has been written within a purely **linguistic** tradition, focusing on subjects of traditional linguistic study and employing what I assume are fundamentally (if not exclusively) linguistic methods of inquiry and argumentation. It has therefore been particularly gratifying to me to discover that the results of my study have useful implications for a number of disciplines apart from theoretical linguistics per se. Thus, it was not my intent to address such lexicographic issues as redundancy in the lexicon, defining formulae for nonpredicating adjectives, arbitrary choices within the definitions of allegedly polysemous adjectives, and associated lexical entries, but, as Section 6.2 has shown, the findings of this study do in fact speak to all these issues.

Somewhat similarly, a number of questions raised in these pages turn out to be of intrinsic interest to psychologists and psycholinguists. These include the question of whether differences in relative productivity among members of the RDP set may be explainable in terms of differences in psychological salience (e.g., are Location and Possession more fundamental relationships for humans to perceive and encode than, say, Causation or Source?), and numerous questions of interest to developmental psycholinguists concerning the ways in which children acquire the rules for productive and creative formation of CNs in their respective languages. Finally, it appears that the semantic and syntactic analysis elaborated in this work has direct relevance to artificial intelligence studies concerned with recognition processes in speech understanding. The contributions of this book that seem most pertinent to research in this area include (*a*) the association of a limited and specifiable set of semantic structures with any given CN, (*b*) the possibility for reducing the inherent ambiguity by means of semantic and pragmatic principles, and (*c*) the syntactic argumentation that shows the internal structure of all CNs to be a series of binary adjunctions of simple nouns and the syntactic category of each CN as a whole to be a noun as well.

The preceding paragraphs are, of course, in no way intended to suggest that the more obviously "linguistic" questions have all been given satisfactory answers in the preceding chapters. On the contrary, enough perplexing and challenging problems remain to satisfy a whole band of future dissertation writers. To begin with, we certainly ought to ask whether the specific analysis of the RDP set which is provided in Chapter 4 is really the best we can do, or whether the evidence demands that important additions—or even conflations—be made in this grammatically crucial set. (This assumes that the notion of an RDP set is itself ultimately defensible, an assumption which is of course also open to challenge.) Next, the origin and role of determiners in the formation of CNs must be analyzed, and the results integrated into the derivations of Chapters 4 and 5. Still another puzzle whose solution remains to be found is why the processes of CN formation create anaphoric islands out of the first element of a CN, while leaving free access to the adjacent head noun for both inbound and outbound anaphora.

One of the broader questions raised but not answered here concerns the appropriate treatment in linguistic theory of word formation processes. Evidence has been given in a number of places suggesting that CN formation shares some of the characteristics of syntactic rules and processes (e.g., cyclicity) and some of the features more commonly associated with morphological or lexical phenomena (e.g., degrees of productivity). The question which then arises is whether our present linguistic framework(s) can accommodate a set of processes that apparently straddle the distinction traditionally drawn between "syntactic" and "morphological" phenomena. A great deal of careful study of a variety of morphologically dissimilar languages is likely to be necessary before this particular question comes close to being answered.

                          8 SUMMARY AND CONCLUSIONS

It is surely obvious by now that although this book may be taken as a sign that some progress has been achieved in understanding complex nominals, the end of this particular road is not yet in sight. This is, of course, not because linguists have been less than diligent in studying CNs but rather because there simply **are** no ends to scholarly roads such as this one. Instead, the territory that we set out to explore is somehow perpetually enlarged and extended by the very act of our exploration. Thus, regardless of how furiously we work in tracking down answers to questions that may have troubled other scholars as well as ourselves, we are always tripping over new questions or even serendipitously creating them out of our own mental flurries.

I have therefore tried to present as fair and complete an account as possible not only of the answers proposed here for "old" questions but also of the myriad of new and intriguing questions that have naturally arisen in the course of this inquiry. If at the same time I have succeeded in conveying to the reader some measure of the challenge, complexity, and satisfactions that I have found in exploring this rich territory, my purpose in writing this book will be amply fulfilled.

APPENDIX

Classification of Complex Nominals
by Proposed Derivations

The lists of CN forms which follow are arranged according to the
specific derivations which I claim are most appropriate to each form when
used to denote its customary referent. Groupings A through L represent
CNs derived by predicate deletion, as analyzed in Chapter 4; detailed
discussion of each of these groupings may be found in Section 4.1.3, and
full derivations in Section 4.2. Groupings M through P represent CNs
derived by predicate nominalization, as analyzed in Chapter 5; discussion
of each nominalization type may be found in Section 5.1, with full
derivations following in Section 5.3.

The subgroupings shown within the predicate deletion categories
(i.e., in Groups F, H, I, K, and L) are intended simply to indicate certain
obvious semantic classes which appear with some frequency within the
derivational pattern in question; these subgroupings are of no syntactic
significance, however, and are irrelevant to the transformational derivations of the larger group. In contrast, the subgroupings shown within the
predicate nominalization categories (i.e., in Groups M and N) are syntactically significant in that they reflect distinct sources of the prenominal
modifier. (See Chapter 5 for more detail.)

To facilitate interpretation of these listings, each derivational pattern is illustrated predominantly, if not exclusively, by CN forms with relatively familiar referents; in this way, the need to provide glosses for each of the many examples is obviated. As a consequence, however, the reader must be careful to keep in mind that the classification of these forms shown below must be understood solely as an indication as to which derivation is most appropriate for the particular **pairing** of a certain form with its customary referent. Such listings in no way imply that the CN form in question has been, or can be, assigned a unique derivational path; indeed, as has been argued throughout this text, every CN form has a number of possible semantic sources, and hence, a number of possible derivations.

To illustrate, let us consider the form *music box* which is listed below under MAKE₁. This classification implies that this CN, when used to denote that entertaining little gadget which is its **customary** referent, is derived from a semantic structure corresponding to *box which makes music* by a number of steps which include MAKE Deletion. This listing does not imply, however, that this is the only way in which *music box* might be derived. For example, if we wished to derive the same form when used to denote an object whose purpose is to hold music (along the lines of *tool box* or *pencil box* on **their** customary readings), the appropriate listing for that pairing of form and meaning would be within the category of CNs derived by deletion of the predicate FOR.

As a result of the inherent ambiguity of CN forms in general, as well as of the kinds of individual variation in interpretation and usage discussed earlier (Section 7.3), the reader may well disagree with the placement of one or another form in the listings below. Such disagreements are to be expected, and do not materially affect the theoretical claims inherent in the classification presented here. Rather, what one must ask oneself in perusing these lists is whether or not the examples shown, together with any others that may come to mind, support the contention that each of these **categories** represents a productive derivational pattern by means of which familiar and—more importantly—novel CN forms are produced and processed by speakers of English. It is this much broader question which is of primary interest here.

A. CAUSE₁

accident weather, concussion force, disease germ, flu virus, growth hormone, sob story, suspense film, tear gas, malarial mosquitoes, mortal blow, traumatic event, whistleberries [euphemistic slang for baked beans]

B. CAUSE$_2$

birth pains, drug deaths, electric shock, future shock, heat rash, thermal cracks, laugh wrinkles, job tension, onion tears, occupational hazard, pot high, snow blindness, moth hole, water mark, hysterical paralysis, traumatic arthritis [technical term for swelling of the joints caused by a specific physical trauma], *cigarette burn, deficiency disease, air pressure, vapor lock, fatigue headache, coffee nerves, genetic disease*

C. HAVE$_1$

picture book, apple cake, gunboat, extension ladder, color television, pictorial atlas, musical comedy, salt lake, fruit tree, vegetable soup, industrial area, cream sauce, college town, nut bread, pet families, coriander curry, lace handkerchief, bear country

D. HAVE$_2$

student power, lemon peel, government land, family antiques, fish scales, apple core, people power, pole height, feminine intuition, presidential power, judicial discretion, professional standards, municipal property, imperial bearing, maternal instinct, planetary mass, vocal range, national resources, company assets, parental prerogatives, city wall, tire rim, student problems, national capital, party members, enemy strength

E. MAKE$_1$

honey bee, silkworm, song bird, sap tree, music box, musical clock, salivary glands, thermal underwear (?)

F. MAKE$_2$

   a. Count-noun modifiers, configurational heads:
     daisy chains, molecular patterns, stellar configurations, floral wreath, acidic crystals, consonantal roots, nervous system, cable network, cellblock, mountain range

   b. Mass-noun modifiers denoting constituent material, heads denoting artifacts or masses:[1]
     chocolate bar, snowball, sugar cube, meatballs, water drop, sand dune, stone tools, bronze statue, copper coins, paper money, concrete desert, steel helmet, plastic toys, candy cigarette, glass eye

   c. Human collectives:
     warrior caste, student committee, worker teams, immigrant minority

---

[1] This entire group has a regular alternative analysis under BE; see Section 7.3 for a discussion of the analytic indeterminacy exemplified by this group.

G.   USE

steam iron, pressure cooker, manual shift, hand brake, electrical clock, hydrogen bomb, water wheel, voice vote, nuclear weapons, solar generator, chemical engineering, smoke signals, machine translation, psychological warfare, faith cure, radio communication, shock treatment, windmill, milieu therapy, needlework, telephonic messages, affixal negation, starvation diet, gas stove, vehicular transport, finger cymbals, vacuum cleaner, vocal music

H.   BE

a. Compositional type (alternative analysis under MAKE₂):
chocolate bar, snowball, bronze statue, copper coins [and all others listed under MAKE₂ in Group Fb above]

b. Genus–species type:
pine tree, mammalian vertebrates, flounder fish, cactus plant, collie dog, protozoic organisms, winter season, sports activities, cash basis, murder charge, consonantal segments, vocalic morphemes, pedal extremities, teaching profession, heart design, star shape, textbook, acidic solvent, aquatic habitat, pictorial evidence

c. Metaphorical type:
sister node, parent organization, ceiling price, head noun, queen bee, soldier ant, mother church, satellite nation, finger lakes, infant colonies, blanket excuse, garter snake, phantom limb, wastebasket category, frog man, hermit crab, target structure, handlebar mustache, beehive hairdo, hairpin turn, pet theory

d. Other (Groups b and d may overlap):
clerical enemies, student friends, reptilian pets, rabbinical judges, sentential subjects, canine companion, women professors, lion cub, servant girl, tape measure, novelty item, parental visitors, vehicular obstruction, child actor, citizen soldier

I.   IN

a. Concrete location type:
marginal note, marine life, field mouse, desert rat, city folk, surface tension, neighborhood bars, urban riots, office friendships, terrestrial life, aquatic mammals, spinal inflammation, mountain lodge

b. Abstract location type:
government employment, theological fallacy, logical impossibility, communist tenet, industrial unrest, family problems, movement schisms, linguistic difficulties, marital conflicts

c. Temporal type:
autumnal rains, nocturnal flight, morning prayers, summer travels, childhood dreams, adolescent turmoil, midnight snack, summer months, evening hours, winter sports, night flight, weekend boredom

J. FOR

avian sanctuary, aldermanic salary, industrial equipment, doghouse, administrative office, cooking utensils, nose drops, picture album, headache pills, horse doctor, arms budget, nasal mist, bull ring, instructional materials, basketball season, digestive system, mothballs, pet spray, plant food, tropical clothing, mining engineer, lightning rod, oil well, Coke machine, automobile plant, juvenile court, sanitary engineering, choral night

K. FROM

a. Products or by-products from natural sources:
olive oil, cane sugar, grain alcohol, rye whiskey, peanut butter, tobacco ash, wood shavings, coal dust, pork suet, bacon grease, alligator leather, rice paper, plum wine
b. Modifier denotes previous location and/or place of origin:
store clothes, home remedy, testtube baby, kennel puppies, country butter, country visitors, sea breeze, farm boy (?), gutter language

L. ABOUT

a. Speech act type (the act, its result, or its record):
abortion vote, procedural motion, criminal policy, tax law, history conference, financial report, theoretical claims, linguistic journal, commercial agreement, political discussion, budget speech, morphology lecture, policy matters, ideological debate
b. Representational type:
pastoral art, disaster flick, adventure story, sports magazine, political cartoons, social satire, academic novel, historical drama, love song
c. "Crisis and conflict" type:
oil crisis, cigarette war, price dispute, sex scandal, energy emergency, constitutional confrontation, abortion problem, financing dilemma, language riots

M. ACT Nominalizations

a. Subjective (modifier < subject NP):
parental refusal, staff attempts, enemy invasion, cell division,

judicial betrayal, student decision, papal vow, animal attack, ship landing, bird reproduction, police intervention

   b. Objective (modifier < direct object NP):
dream analysis, birth control, musical criticism, heart massage, subject deletion, child abuse, jungle exploration, land reclamation, automotive repair, acoustic amplification, urban planning

N.   PRODUCT Nominalizations

   a. Subjective (modifier < subject NP):
clerical errors, editorial comments, royal orders, senatorial nominations, peer judgments, faculty decisions, papal appeals, presidential appointments, national exports, student critiques

   b. Objective (modifier < direct object NP):
musical critiques, oil imports, constitutional amendment, tuition subsidies, food supplies, liquor orders, book requests, vocalic insertions, cover designs, beard trim, artifact descriptions

O.   AGENT Nominalizations

city planner, financial analyst, artistic manager, literary editor, acoustic amplifier, film cutter, mail sorter, blood donor, census taker, orchestral conductor, draft dodger, metallic separator

P.   PATIENT Nominalizations

student inventions, designer creations, royal gifts, national requirements, mammalian secretions, parental requests, factory rejects, national imports, city trainees, senatorial nominees, presidential appointees, college employees

REFERENCES

Adams, Valerie (1973). *An introduction to modern English word formation*. London: Longman.

Akmajian, Adrian (1975). More evidence for an NP cycle. *Linguistic Inquiry* 6:115–129.

Allerton, D. J. (1972). Review of H. E. Brekle (1970), *Generative Satzsemantik und transformationelle Syntax im System der Englischen Nominalkomposition* (Munich: W. Fink) and S. Žepić (1970), *Morphologie und Semantik der deutschen Nominalkomposita* (Zagreb: Philosophische Fakultät der Universität). *Journal of Linguistics* 8:321–326.

Anderson, Stephen R. (1976). Concerning the notion "base component of a transformational grammar." In James D. McCawley (Ed.), *Syntax and semantics, Volume 7: Notes from the linguistic underground*. New York: Academic Press. Pp. 113–128.

Apresyan, Yu. D., I. A. Mel'čuk, and A. K. Žolkovsky (1970). Semantics and lexicography: Towards a new type of unilingual dictionary. In F. Kiefer (Ed.), *Studies in syntax and semantics*. Dordrecht, Holland: D. Reidel. Pp. 1–33.

Aronoff, Mark (1976). *Word formation in generative grammar*. Linguistic Inquiry Monograph One. Cambridge, Massachusetts: The MIT Press.

Bach, Emmon (1967). *Have* and *be* in English syntax. *Language* 43:462–485.

Bach, Emmon (1968). Nouns and noun phrases. In Emmon Bach and Robert T. Harms (Eds.), *Universals in linguistic theory*. New York: Holt, Rinehart and Winston, Inc. Pp. 91–122.

Barbaud, Philippe (1971). L'ambiguïté structurale du composé binominal. *Cahier de linguistique N° 1*. Montreal: Les Presses de l'Université du Québec. Pp. 71–116.

Bartning, Inge (1976). *Remarques sur la syntaxe et la sémantique des pseudo-adjectifs dénominaux en français*. Stockholm: Institut d'Etudes Romanes, Université de Stockholm.

Becker, A. L. and D. G. Arms (1969). Prepositions as predicates. In Robert I. Binnick, Alice Davison, Georgia M. Green, and Jerry L. Morgan (Eds.), *Papers from the fifth regional meeting: Chicago Linguistic Society*. Chicago: Chicago Linguistic Society. Pp. 1–11.

Berman, Arlene (1974). Adjectives and adjective complement constructions in English. Formal Linguistics Report No. NSF-29 to the National Science Foundation. Cambridge, Massachusetts: Harvard University.

Binnick, Robert I. (1970). Studies in the derivation of predicative structures. *Papers in Linguistics* 3:237–340, 519–602.

Bolinger, Dwight (1967). Adjectives in English: Attribution and predication. *Lingua* 18:1–34.

Bolinger, Dwight (1972). *Degree words*. The Hague: Mouton.

Borkin, Ann (1972). Coreference and beheaded NP's. *Papers in Linguistics* 5:28–45.

Botha, Rudolf P. (1968). *The function of the lexicon in transformational generative grammar*. Janua Linguarum, Series Maior, 38. The Hague: Mouton.

Brekle, H. E. (1970). *Generative Satzsemantik und transformationelle Syntax im System der Englischen Nominalkomposition*. Munich: W. Fink.

Brugmann, K. (1900). Über das Wesen der sogenannten Wortzusammensetzungen. Eine sprachpsychologische Studie. *Berichte über die Verhandlungen der königlichen sächsischen Gesellschaft der Wissenschaften zu Leipzig*, Phil.-Hist. Classe 52:359–401.

Burt, Marina (1971). *From deep to surface structure*. New York: Harper & Row.

Chomsky, Noam (1965). *Aspects of the theory of syntax*. Cambridge, Massachusetts: The MIT Press.

Chomsky, Noam (1970). Remarks on nominalizations. In Roderick A. Jacobs and Peter S. Rosenbaum (Eds.), *Readings in English transformational grammar*. Waltham, Massachusetts: Ginn and Company. Pp. 184–221.

Chomsky, Noam (1972a). Some empirical issues in the theory of transformational grammar. In Stanley Peters (Ed.), *Goals of linguistic theory*. Englewood Cliffs, New Jersey: Prentice-Hall, Inc. Pp. 63–130.

Chomsky, Noam (1972b). *Studies on semantics in generative grammar*. The Hague: Mouton.

Chomsky, Noam and Morris Halle (1968). *The sound pattern of English*. New York: Harper & Row.

Chryst, Joseph Clarc (1976). Grammatical ambiguity in Japanese and English. Unpublished master's thesis, University of Northern Iowa.

Clancy, Patricia (1975). An experiment in compound formation. Unpublished manuscript, University of California at Berkeley.

Coates, Jennifer (1971). Denominal adjectives: A study in syntactic relationships between modifier and head. *Lingua* 27:160–169.

Coursaget-Colmerauer, Colette (1973). Les déterminants de la nominalisation. *Cahier de linguistique N° 3*. Montreal: Les Presses de l'Université du Québec. Pp. 39–51.

Crystal, D. and D. Davy (1969). *Investigating English style*. London: Longmans.

Davidson, Donald (1967). The logical form of action sentences. In Nicholas Rescher (Ed.), *The logic of decision and action*. Pittsburgh: University of Pittsburgh Press. Pp. 81–95.

Dede, Müşerref (1977). Where do Turkish nominal compounds come from? Unpublished manuscript, University of Michigan.

Dell, F. (1970). Les règles phonologiques tardives et la morphologie dérivationelle du français. Unpublished doctoral dissertation, Massachusetts Institute of Technology.

Dixon, R. M. W. (1970). Where have all the adjectives gone? An essay in universal semantics. Unpublished manuscript, Australian National University, Canberra. [Reprinted in *Studies in Language* 1:19–80, 1977.]

Downing, Pamela Ann (1975). Pragmatic constraints on nominal compounding in English. Unpublished master's thesis, University of California at Berkeley.

Downing, Pamela Ann (1977). On the creation and use of English compound nouns. *Language* 53:810–842.

Fraser, Bruce (1970). Some remarks on the action nominalization in English. In Roderick A. Jacobs and Peter S. Rosenbaum (Eds.), *Readings in English transformational grammar*. Waltham, Massachusetts: Ginn and Company. Pp. 83–98.

Geis, Michael L. (1970). Time prepositions as underlying verbs. In Mary Ann Campbell *et al.* (Eds.), *Papers from the sixth regional meeting: Chicago Linguistic Society*. Chicago: Chicago Linguistic Society. Pp. 235–249.

Givón, Talmy (1970). Notes on the semantic structure of English adjectives. *Language* 46:816–837.

Gleitman, Lila R. and Henry Gleitman (1970). *Phrase and paraphrase: Some innovative uses of language*. New York: W. W. Norton.

Green, Georgia M. (1974). *Semantics and syntactic regularity*. Bloomington: Indiana University Press.

Grice, H. Paul (1975). Logic and conversation. In Peter Cole and Jerry L. Morgan (Eds.), *Syntax and semantics 3: Speech acts*. New York: Academic Press. Pp. 41–58.

Gruber, Jeffrey S. (1965). Studies in lexical relations. Unpublished doctoral dissertation, Massachusetts Institute of Technology. Distributed by Indiana University Linguistics Club. [Reprinted in Jeffrey S. Gruber, *Lexical structures in syntax and semantics*. Amsterdam: North-Holland, 1976.]

Gruber, Jeffrey S. (1967). Functions of the lexicon in formal descriptive grammars. Technical memorandum series, Santa Monica: Systems Development Corporation. [Reprinted in Jeffrey S. Gruber, *Lexical structures in syntax and semantics*. Amsterdam: North-Holland, 1976.]

Hall, Robert A., Jr. (1973). The transferred epithet in P. G. Wodehouse. *Linguistic Inquiry* 4:92–94.

Hankamer, Jorge (1973). Unacceptable ambiguity. *Linguistic Inquiry* 4:17–68.

Hatcher, Anna Granville (1960). An introduction to the analysis of English noun compounds. *Word* 16:356–373.

Jackendoff, Ray (1975). Morphological and semantic regularities in the lexicon. *Language* 51:639–671.

Jackendoff, Ray (1977). $\overline{X}$ *syntax: A study of phrase structure*. Linguistic Inquiry Monograph 2. Cambridge, Massachusetts: The MIT Press.

Jacobs, Roderick A. and Peter S. Rosenbaum (1968). *English transformational grammar*. Waltham, Massachusetts: Ginn and Company.

Jenkins, Janet C. (1976). An investigation of the beheading analysis for metonymous NP's. Unpublished manuscript, Northwestern University.

Jespersen, Otto (1942). *A modern English grammar on historical principles*. Copenhagen: Ejnar Munksgaard.

Joos, Martin (1961). *The five clocks*. New York: Harcourt, Brace and World.

Kauffmann, Stanley (1974). Review of *Jumpers*, by Tom Stoppard. *New Republic* 18 May 1974, p. 18.

Kay, Paul and Karl Zimmer (1976). On the semantics of compounds and genitives in English. Paper presented at the Sixth Annual Meeting of the California Linguistics Association, San Diego, California, May 1976.

Koziol, Herbert (1937). *Handbuch der englischen Wortbildungslehre*. Heidelberg: Carl Winter's Universitätsbuchhandlung.

Labov, William (1972). *Sociolinguistic patterns*. Philadelphia: University of Pennsylvania Press.

Lakoff, George (1970). *Irregularity in syntax*. New York: Holt, Rinehart and Winston.

Lakoff, George (1972). Hedges: A study in meaning criteria and the logic of fuzzy concepts. In Paul M. Peranteau, Judith N. Levi, and Gloria C. Phares (Eds.), *Papers from the eighth regional meeting: Chicago Linguistic Society*. Chicago: Chicago Linguistic Society. Pp. 183–228.

Lakoff, George and Stanley Peters (1969). Phrasal conjunction and symmetric predicates. In David A. Reibel and Sanford A. Schane (Eds.), *Modern studies in English*. Englewood Cliffs, New Jersey: Prentice-Hall. Pp. 113–142.

Leech, Geoffrey (1974). *Semantics*. Harmondsworth, England: Penguin Books.

Lees, Robert B. (1960). *The grammar of English nominalizations*. International Journal of American Linguistics 26, Publication 12.

Lees, Robert B. (1966). On a transformational analysis of compounds: A reply to Hans Marchand. *Indogermanische Forschungen* 71:1–13.

Lees, Robert B. (1970). Problems in the grammatical analysis of English nominal compounds. In Manfred Bierwisch and Karl Erich Heidolph (Eds.), *Progress in linguistics*. The Hague: Mouton. Pp. 174–186.

Levi, Judith N. (1973). Where do all those other adjectives come from? In Claudia Corum, T. Cedric Smith-Stark, and Ann Weiser (Eds.), *Papers from the ninth regional meeting: Chicago Linguistic Society*. Chicago: Chicago Linguistic Society. Pp. 332–345.

Levi, Judith N. (1974). On the alleged idiosyncrasy of nonpredicate NPs. In Michael W. LaGaly, Robert A. Fox, and Anthony Bruck (Eds.), *Papers from the tenth regional meeting: Chicago Linguistic Society*. Chicago: Chicago Linguistic Society. Pp. 402–415.

Levi, Judith N. (1975). The syntax and semantics of nonpredicating adjectives in English. Unpublished doctoral dissertation, The University of Chicago. Distributed by the Indiana University Linguistics Club.

Levi, Judith N. (1976). A semantic analysis of Hebrew compound nominals. In Peter Cole (Ed.), *Studies in Modern Hebrew syntax and semantics: A transformational–generative approach*. Amsterdam: North-Holland. Pp. 9–55.

Levi, Judith N. (1977). The constituent structure of complex nominals, or, That's funny, you don't *look* like a noun! In Woodford A. Beach, Samuel E. Fox, and Shulamith Philosoph (Eds.), *Papers from the thirteenth regional meeting: Chicago Linguistic Society*. Chicago: Chicago Linguistic Society. Pp. 325–338.

Li, Charles N. (1971). Semantics and the structure of compounds in Chinese. Unpublished doctoral dissertation, University of California at Berkeley.

Livant, William Paul (1962). Productive grammatical operations: I: The noun compounding of 5-year-olds. *Language Learning* 12:15–26.

Ljung, Magnus (1970). *English denominal adjectives: A generative study of the semantics of a group of high-frequency denominal adjectives in English*. Gothenburg Studies in English. Lund, Sweden: University of Gothenburg.

Marchand, Hans (1969). *The categories and types of present-day English word formation*, second edition. Munich: C. H. Beck'sche Verlagsbuchhandlung.

Mardirussian, Galust (1975). Noun-incorporation in universal grammar. In Robin E. Grossman, L. James San, and Timothy J. Vance (Eds.), *Papers from the eleventh regional meeting: Chicago Linguistic Society*. Chicago: Chicago Linguistic Society. Pp. 383–389.

Markowitz, Judith A. (1977). A look at fuzzy categories. Unpublished doctoral dissertation, Northwestern University.

Matthews, P. H. (1974). *Morphology: An introduction to the theory of word-structure*. Cambridge: Cambridge University Press.

McCawley, James D. (1968). Lexical insertion in a transformational grammar without deep structure. In Bill J. Darden, Charles-James N. Bailey, and Alice Davison (Eds.), *Papers from the fourth regional meeting: Chicago Linguistic Society*. Chicago: Chicago Linguistic Society. Pp. 71–93.

McCawley, James D. (1970). Where do noun phrases come from? In Roderick A. Jacobs and Peter S. Rosenbaum (Eds.), *Readings in English transformational grammar*. Waltham, Massachusetts: Ginn and Company. Pp. 166–183.

McCawley, James D. (1973). External NPs versus annotated deep structures. *Linguistic Inquiry* 4:221–240.

McCawley, James D. (1974). On identifying the remains of deceased clauses. Unpublished manuscript, University of Chicago. [Reprinted in Diane D. Bornstein (Ed.), *Readings in the theory of grammar*. Cambridge, Massachusetts: Winthrop. Pp. 288–298.]

McCawley, James D. (1975). Review of Noam Chomsky (1972), *Studies on semantics in generative grammar* (The Hague: Mouton). *Studies in English Linguistics* [Japan] 3:209–311. [Earlier version distributed in mimeographed form by Indiana University Linguistics Club.]

McCawley, James D. (1976a). Morphological indeterminacy in underlying syntactic structure. In Frances Ingemann (Ed.), *1975 Mid-America Linguistics Conference papers*. Lawrence, Kansas: University of Kansas Linguistics Department. Pp. 317–326.

McCawley, James D. (1976b). Notes on Jackendoff's theory of anaphora. *Linguistic Inquiry* 7:319–341.

McCawley, James D. (1976c). Remarks on what can cause what. In Masayoshi Shibatani (Ed.), *Syntax and semantics 6: The grammar of causative constructions*. New York: Academic Press. Pp. 117–129.

McCawley, James D. (1976d). Some ideas not to live by. *Die neueren Sprachen* 75:151–165.

Meys, W. J. (1975). *Compound adjectives in English and the ideal speaker-listener*. North-Holland Linguistic Series 18. Amsterdam: North-Holland.

Motsch, Wolfgang (1970). Analyse von Komposita mit zwei nominalen Elementen. In Manfred Bierwisch and Karl Erich Heidolph (Eds.), *Progress in linguistics*. Janua Linguarum, Series Maior 43. The Hague: Mouton. Pp. 208–223.

Newmeyer, Frederick J. (1970). The derivation of the English action nominalization. In Mary Ann Campbell *et al.* (Eds.), *Papers from the sixth regional meeting: Chicago Linguistic Society*. Chicago: Chicago Linguistic Society. Pp. 408–415.

Newmeyer, Frederick J. (1971). The source of derived nominals in English. *Language* 47:786–796.

Newmeyer, Frederick J. (1975). The position of incorporation transformations in the grammar. In Cathy Cogen *et al.* (Eds.), *Proceedings of the first annual meeting of the Berkeley Linguistics Society*. Berkeley: Berkeley Linguistics Society. Pp. 333–342.

Newmeyer, Frederick J. (1976). The precyclic nature of predicate raising. In Masayoshi Shibatani (Ed.), *Syntax and semantics 6: The grammar of causative constructions*. New York: Academic Press. Pp. 131–163.

Oestman, Bethel (1972). Interpretation of compound nouns: Children vs. adults. Unpublished manuscript, University of California at Berkeley.

Postal, Paul M. (1969). Anaphoric islands. In Robert I. Binnick, Alice Davison, Georgia M. Green, and Jerry L. Morgan (Eds.), *Papers from the fifth regional meeting: Chicago Linguistic Society*. Chicago: Chicago Linguistic Society. Pp. 205–239.

Postal, Paul M. (1970). On the surface verb "remind." *Linguistic Inquiry* 1:57–120.

Postal, Paul M. (1972). The derivation of English pseudo-adjectives. Unpublished manuscript, Thomas J. Watson Research Center, IBM (Yorktown Heights, New York).

Postal, Paul M. (1974). *On raising: One rule of English grammar and its theoretical implications*. Current Studies in Linguistics Series. Cambridge, Massachusetts: The MIT Press.

Reddy, Michael J. (1969). A semantic approach to metaphor. In Robert I. Binnick, Alice Davison, Georgia M. Green, and Jerry L. Morgan (Eds.), *Papers from the fifth regional meeting: Chicago Linguistic Society*. Chicago: Chicago Linguistic Society. Pp. 240–251.

Reddy, Michael J. (1973). Formal referential models of poetic structure. In Claudia Corum, T. Cedric Smith-Stark, and Ann Weiser (Eds.), *Papers from the ninth regional meeting: Chicago Linguistic Society*. Chicago: Chicago Linguistic Society. Pp. 493–518.

Reddy, Michael J. (1977). The conduit metaphor: A case of frame conflict in our language about language. Paper read at Interdisciplinary Conference on Metaphor and Thought, University of Illinois at Urbana-Champaign, September 29, 1977.

Reif, Joseph A. (1968). Construct state nominalizations in Modern Hebrew. Unpublished doctoral dissertation, University of Pennsylvania.

Rips, L. J., E. J. Shoben, and E. E. Smith (1973). Semantic distance and the verification of semantic relations. *Journal of Verbal Learning and Verbal Behavior* 12:1–20.

Rosch, Eleanor (1973). On the internal structure of perceptual and semantic categories. In T. Moore (Ed.), *Cognitive development and the acquisition of language*. New York: Academic Press. Pp. 111–144.

Ross, John Robert (1967). Constraints on variables in syntax. Unpublished doctoral dissertation, Massachusetts Institute of Technology. Distributed by Indiana University Linguistics Club.

Ross, John Robert (1969). A proposed rule of tree pruning. In David A. Reibel and Sanford A. Schane (Eds.), *Modern studies in English*. Engelwood Cliffs, New Jersey: Prentice-Hall, Inc. Pp. 288–299.

Ross, John Robert (1972a). The category squish: Endstation Hauptwort. In Paul M. Peranteau, Judith N. Levi, and Gloria C. Phares (Eds.), *Papers from the eighth regional meeting: Chicago Linguistic Society*. Chicago: Chicago Linguistic Society. Pp. 316–328.

Ross, John Robert (1972b). Act. In Donald Davidson and Gilbert Harman (Eds.), *Semantics of natural language*. Dordrecht, Holland: D. Reidel. Pp. 70–126.

Ross, John Robert (1974a). Nouniness. In Osamu Fujimura (Ed.), *Three dimensions of linguistic theory*. Tokyo: TEC Corporation. Pp. 137–257.

Ross, John Robert (1974b). To have *have* or to not have *have*. Unpublished manuscript, Massachusetts Institute of Technology.

Sadock, Jerrold M. (1974). Read at your own risk: Syntactic and semantic horrors you can find in your own medicine chest. In Michael W. LaGaly, Robert A. Fox, and Anthony Bruck (Eds.), *Papers from the tenth regional meeting: Chicago Linguistic Society*. Chicago: Chicago Linguistic Society. Pp. 599–607.

Sadock, Jerrold M. and Judith N. Levi (1977). Ergativity in English? In Samuel E. Fox, Woodford A. Beach, and Shulamith Philosoph (Eds.), *The CLS book of squibs with cumulative index*. Chicago: Chicago Linguistic Society. Pp. 91–93.

Schachter, Paul (1977). Constraints on coordination. *Language* 53:86–103.

Selkirk, Elizabeth (1977). Some remarks on noun phrase structure. In Peter W. Culicover, Thomas Wasow, and Adrian Akmajian (Eds.), *Formal syntax: Proceedings of the MSSB–UC Irvine conference on the formal syntax of natural language*. New York: Academic Press. Pp. 285–316.

Shibatani, Masayoshi (1973). A linguistic study of causative constructions. Unpublished doctoral dissertation, University of California at Berkeley.

Shibatani, Masayoshi (1976). (Ed.) *Syntax and semantics 6: The grammar of causative constructions*. New York: Academic Press.

Thompson, Sandra A. (1974). On the issue of productivity in the lexicon. In Sandra A. Thompson and Carol Lord (Eds.), *Approaches to the lexicon*. UCLA Papers in Syntax, No. 6. Los Angeles: University of California at Los Angeles. Pp. 1–25.

Traugott, Elizabeth Closs (1975). Spatial expressions of tense and temporal sequencing: A contribution to the study of semantic fields. *Semiotica* 15:207–230.

Turner, G. W. (1973). *Stylistics*. Harmondsworth, Middlesex: Penguin Books.

Vendler, Zeno (1967). *Linguistics in philosophy*. Ithaca, New York: Cornell University Press.

Vendler, Zeno (1968). *Adjectives and nominalizations*. Papers on Formal Linguistics No. 5. The Hague: Mouton.

Warshawsky Harris, Florence (1965). Reflexivization. Unpublished manuscript, Massachusetts Institute of Technology. [Reprinted in James D. McCawley (Ed.), *Syntax and semantics 7: Notes from the linguistic underground*. New York: Academic Press, 1976. Pp. 63–83.]

Wentworth, Harold and Stuart Berg Flexner (1975). *Dictionary of American slang*, 2nd supplemented edition. New York: Thomas Y. Crowell Company.

Whitney, William Dwight (1879). *A Sanskrit grammar*. Leipzig: Breitkopf und Härtel.

Williams, Edwin S. (1974). Rule ordering in syntax. Unpublished doctoral dissertation, Massachusetts Institute of Technology.

Zadeh, Lofti (1971). *Fuzzy languages and their relation to human and machine intelligence*. Memorandum No. ERL-M302, Electronic Research Laboratory. Berkeley: University of California at Berkeley.

Zimmer, Karl E. (1964). *Affixal negation in English and other languages: An investigation of restricted productivity*. Monograph No. 5, Supplement to *Word* 20:2.

Zimmer, Karl E. (1971). Some general observations about nominal compounds. *Working Papers on Language Universals, Stanford University* 5:C1–C21.

Zimmer, Karl E. (1972). Appropriateness conditions for nominal compounds. *Working Papers on Language Universals, Stanford University* 8:3–20.

Zwicky, Arnold M. and Jerrold M. Sadock (1975). Ambiguity tests and how to fail them. In John Kimball (Ed.), *Syntax and semantics 4*. New York: Academic Press. Pp. 1–36.